The Widening Breach

The Widening

Evolutionism in the

Breach:

Mirror of Cosmology

Whitall N. Perry

QE

This edition published 1995 by
Quinta Essentia, P.O. Box 842
Bartlow, Cambridge, CB1 6PX, UK.

ISBN 1 870196 13 9 *paper*

British Library Cataloguing in Publication Data
A catalogue record of this book is available from
The British Library

Typeset by Goodfellow & Egan, Cambridge
Printed in the United Kingdom

Contents

Preface ix

1 The Missing Link 1

2 *Creatio ex Nihilo* 11

3 Contours of the Primordial Tradition 21

4 Realism to Nominalism: The Watershed 44

5 Nominalism to Atomism: The Outer Darkness 73

6 The Sundoor 96

Suggested Reading 102

Index 104

List of Illustrations

Fig. 1 (Facing page 46) Balinese temple with split gate: the northern temple Meduwe Karang at Kubutambahan. (*Photograph courtesy of Werner Foreman Archive*).

Fig. 2 (Page 23) The eight Chinese Trigrams (schematic cosmic polarities) enclosing Yin-Yang symbol.

Fig. 3 (Between pages 46–47) Emerging from the Night of Beatitude onto the stage of time. Balinese dancer in a troupe performing at the Palais de Beaulieu in Lausanne, September 1957. (*Photograph by the author*).

Fig. 4 (Between pages 46–47) ST. GEORGE. Prototype, like St. Nicholas, of the great apotropaic and intercessionary saints of the early Church.

This icon, designated as 'akheiropoiite' (not made with the hand), is in the monastery of Zographou on Mount Athos having, according to legend, been miraculously transported from an earlier monastery of St. George at Phanuel near Lydda, Palestine, *c*886–911. (*From the collection Orbis Pictus, Vol. 43,* Icônes du Mont Athos, *by Paul Huber, Payot, Lausanne,* 1965).

Fig. 5 (Between pages 46–47) KWANNON, Japanese. (*Photograph taken by author in the Musée de l'Ariana, Geneva*).

Fig. 6 (Page 51) T-O Map, seventh century. From Isodor of Seville's 'Origines'.

Fig. 7 (Between pages 78–79) CHARLEMAGNE, by Dürer. Exemplar of Holy Roman Emperor. (*Photograph courtesy National Gallery of Art, Washington: Rosenwald Collection*).

Fig. 8 (Between pages 78–79) 'MELENCOLIA', by Dürer. Described in text. (*Photograph courtesy National Gallery of Art, Washington: Rosenwald Collection*).

Fig. 9 (Between pages 78–79) Ceramic tilework, French sixteenth-seventeenth century. Described in text. (*Photograph by the author*).

Fig. 10 (Between pages 78–79) Passage from the world of pairs to the World of Unity. Four Lohans and Two Attendents Witnessing the Transfiguration of a Lohan. Zhou Jiehang, Chinese, Song Dynasty, 1180. Colours on silk. (*Museum Collection, courtesy of the Museum of Fine Arts, Boston*).

The Self, without inside, without outside, is entirely a cognitive pleroma (*prajñâna-ghana*).

Bṛihad-Âraṇyaka Upanishad, IV.V.13.[1]

The wall of the Paradise in which Thou, God, dwellest is built of the contradictories.

Nicholas of Cusa, *De vis Dei*, IX.[2]

1. Tr. A. K. Coomaraswamy. 2. Tr. idem.

Preface

INCLUDING but meaning more than just Darwinism, the term evolutionism as used in this book comprehends an existential vision that runs deep and spontaneously in the veins of present-day humanity, even with the generality of those who consider themselves rationally or spiritually opposed to the concept. We are speaking here of the progress and acceleration of the secular mentality—of which transformism is a crucial facet—even when this mentality allies itself with religion.

Science versus religion, evolutionism versus anti-evolutionism—it is all argued, consciously or not, on the premises of a Newtonian universe. Starting from the same common denominator, namely, an *experienced* objective matter-ambience based on the 'obvious'-seeming independence of the physical world, the dialogue is doomed to perpetuate itself without final issue. A contemporary traditionalist scholar drives home this point in asking: 'How often does knowledge really escape the limits afforded by impenetrables or the shape of machines, or for that matter the even more basic sense of physical resistance at the surface of our skin? Atomism, dualism, mechanism, materialism—a feeling in our very bones of solidity and boundary—are endemic to virtually all cognition and all our cognitive assumptions.'[1] The dualities of which our physical world is composed are so artfully 'woven' into a 'seamless garment' that the sustaining 'warp and woof' go unperceived, even though the actuality is little more

1. James S. Cutsinger, *The Form of Transformed Vision – Coleridge and the Knowledge of God*, Mercer University Press, Macon, GA, 1987, p.58.

than that of a passing spectrum of illusions—each in turn made to seem real to those beguiled by the *outward* universe proportionately to their ignorance of its *inward* dimensions. Creationism itself, when the symbolism is lost, becomes the grafting through faith of a fundamentalist literalism onto a secular set of references that takes no account of a truly transcendent or metaphysical element – excepting the more or less indeterminate idea of a Creator.

Apart from whether or not evolutionism is regarded as an aberrational theory, it has its *raison d'être* within the structure of the cosmic moment in which we find ourselves; if not 'the right doctrine at the right time', it is none the less an idea 'whose time has come' in the evolution of our cosmic cycle, and it has to be dealt with accordingly. Originally this book had for title *Evolutionism in the Mirror of Philosophy*, but we were reminded that evolutionism, being only an hypothesis and not a science, is itself a philosophical construction.

★

The principles underlying the play of polarities developed in the following chapters have already been fully enunciated in their different aspects by Frithjof Schuon, René Guénon, and Ananda K. Coomaraswamy. But principles by their very essentiality are capable of ongoing applications which no one author can exhaust; and there are important deductions deriving from the aforementioned principles which merit elaboration.

Those readers unfamiliar with the traditionalist setting that forms the background of this book may find useful the suggested reading listed at the end.

1 The Missing Link

All things are double, one against another, and he hath made nothing defective. *Ecclesiasticus*, XLII.25

And all things we have created by pairs, that haply ye may reflect. *Qur'ân*, LI.49

THE CREDIBILITY of the evolutionist hypothesis on the origin and development of life does indeed suffer from a missing link, but it is in the domain of cosmology, and not paleontology, that the crux of the matter really resides. The link missing is *de facto* the pole 'subject', apart from which the pole 'object' is inconceivable. And yet ironically, were this first pole present in its thesis on origins, then evolutionism would lose the entire rationale for its premises, for the simple reason that in our physical universe an object necessarily presupposes the existence of a subject and vice versa, as will be demonstrated; and this initial polarity transformism along with modern physics cannot admit.

The *Oxford English Dictionary* defines 'subject' *in the use intended here*, as follows:

> 9. *Modern Philos.* More fully *conscious* or *thinking subject*: The mind, as the 'subject' in which ideas inhere; that to which all mental representations or operations are attributed; the thinking or cognizing agent; the self or ego. (Correlative to Object sb.6.)

Webster's *New International Dictionary* (2nd ed.) gives this definition of 'subject':

> That which sustains or is embodied in thought or consciousness; the thinking agent; the mind, ego, or reality of whatever sort, which supports, or assumes the form of, mental operations; as, in the idealistic dictum, 'No object without a *subject*'.

One more definition, from the *Dictionary of Philosophy* (ed. Dagobert D. Runes, New York, Philosophical Library, 1960):

> *In Epistemology*: The subject of knowledge is the individual knower considered either as a pure ego . . . a transcendental ego . . . or an act of awareness.

Returning to the *Oxford English Dictionary*, we find this definition of 'object':

> 6. *Metaph.* A thing or being of which one thinks or has cognition, as correlative to the thinking or knowing *subject*; something external, *or regarded as external*, to the mind; the non-ego as related to, or distinguished from, the ego [2nd emphasis added].

And from *Webster's*:

> In philosophical discussions, *object* usually stands in antithesis to *subject*, *ego*, or *self-consciousness*, and its meaning may be extended to designate the totality of external phenomena, as constituting the notself. Philosophically, *object* is only one term in the *subject-object* relationship, that is, it is never considered apart from some relation to a subject, but in ordinary usage it is often equivalent to *thing*, or physical entity *conceived as totally independent* [last emphasis added].

The *Dictionary of Philosophy* has this to say on the word:

> In the widest sense, object is that towards which consciousness is directed, whether cognitively or conatively.

These foregoing definitions bring out the reciprocity between the two poles subject-object. *Webster's* remark about the 'ordinary usage' of *object* as a 'physical entity conceived as totally independent' has been rendered in italics because this way of seeing things has only become current, and increasingly so, in Western thought over the past some six hundred years. The fact is, there does or could exist no such phenomenon as a purely physical entity that is totally independent. In the whole of manifested creation there is absolutely not to be found a 'thing', concept, principle, law, measure, property, or force that is conceivable apart from its complementary, opposite, contrasting, or contradictory pole.

Call it what you will: the Law of Pairs, the Law of Dualities, the Law of Polarities, the Law of Corollaries, or Law of Opposites, it is by separation that the worlds are created. Everything manifested proclaims its self-insufficiency by testifying existentially to that other half which it is not. Unity is the property of the Next World. If there could be integral unity down here, then this world would already be the Next. And even in principial Reality itself the essential archetype of polarity prevails, without which there could be no distinction between the Absolute and the Infinite, as is shown in the next chapter.

Commencing with the pair subject-object, one can expand the terms endlessly: knower-known, cause-effect, activity-passivity, emptiness-fullness, interiority-exteriority, increase-decrease, greatness-smallness, expansion-contraction, height-depth, front-back, right-left, precedence-sequence, beginning-end, inclusion-exclusion, attraction-repulsion, convexity-concavity, centripetal-centrifugal, heat-cold, light-darkness, sound-silence, positive-negative, verticality-horizontality, masculine-feminine, the cardinal points, virtue-vice, pleasure-pain, good-evil, life-death, time-eternity, and so on *ad infinitum*.

Ralph Waldo Emerson in his essay 'Compensation' writes:

> Polarity, or action and reaction, we meet in every part of nature; in darkness and light; in heat and cold; in the ebb and flow of waters; in male and female; . . . in the systole and diastole of the heart . . .. An inevitable dualism bisects nature, so that each thing is a half, and suggests another thing to make it whole; as, spirit, matter; man, woman; odd, even; subjective, objective; in, out; upper, under; motion, rest; yea, nay.

★

The point of all this is to ask simply, why should the pair subject-object alone, on the plane of manifested existence not be a 'pair', but be free from the 'tyranny' of interdependence or linkage to which all the other listed and unlisted terms without exception are subjected?

For precisely what the evolutionists are telling us in so many words is that inanimate matter as *object* can 'enjoy' indefinitely, and hence presumably perpetually if it so 'wishes', perfect autonomy with no corollary reference whatsoever to a *subject*, namely, a 'cognizing agent' either principial or manifested. Yet whoso says *known* says *knower*: a galaxy 'out there' is known to be so uniquely by the observation and testimony of a knower 'in here'. Even the 'unknown' is the knower's concept. It is not that evolutionists perforce *think* in terms of pure object, but that they fail to consider the necessity of any pre-existent reality beyond that of physical matter.

They claim that primal 'cosmic dust', thanks to the stimulus of 'friction', arrived over some billions of years at a tropismatic molecular organization of the amino acid constituents of protein, giving us the biochemical components of protoplasm. They would have, among other theories, filter-passing viruses act (through a purely imaginary, unverified process of primordial archebiosis or abiogenesis—life evolving spontaneously from inanimate matter, or the cognized object creating the cognizing subject, when in reality both originate from a primary source)—as the 'missing link' between inanimate and animate matter, with bacteria as the immediate ancestors of protozoa, springing from carbon compounds of protein.

Whether in fact these theories concerning the origin of living cells hold sway still, or whether others have come to replace them, is an affair of little importance. What is of cardinal importance is to understand that no prerogative, cosmic principle or law exists by which this inanimate and subjectless—hence limited—pristine stuff could from its inception maintain over measureless time a perfect self-containment—a self-containment which being *ipso facto* 'absolute' (but then how can it be integrally and hermetically inanimate, since the Absolute obviously includes the animate as part of its Absoluteness?) would make the necessity of producing a subject purely superfluous. And if somehow, in spite of all, an embryonic subject was fused into the original atoms of the cosmos, then it is for any proponents of this theory to enlighten us on the identity of their alleged subject. The long and short of it is that if the *particular* matter that is in question here can be classified neither as a subject nor an object with regard to any context whatsoever, then it enjoys an unlimited autonomy—to which nothing can be added and from which nothing abstracted—an hypothesis contradicted *a priori* by the admission of this matter's *particularity*. The truth is, there is no autonomy in space, time, measure, name, or form.

★

Our use of the term *subject* throughout this book will be primarily that of the dictionary definitions given on pages 1 and 2, namely, that of 'cognizing agent', or the pole consciousness, awareness, perception, which pole in a general sense is considered as being internal or inward. Similarly, our use of the term *object* will be based on the dictionary definitions given on page 2, namely, that of cognized thing (or being), 'regarded as external' to the cognizing agent, or the pole physical matter, entity, thing, which pole in a general sense is considered as being external or outward. The distinction subject-object intended here is thus basically that of self from notself, or more precisely, conscious existence from insentient and inanimate matter.

It is not being claimed that evolutionists think in this way; on the contrary, it is being said that they think in terms of the material or physical pole alone, failing to take into account the indispensable corollary of the pole *subject*.

For the sake of completeness we include in résumé the essential metaphysical formulations which Frithjof Schuon has provided concerning the terms Subject and Object:

The absolute Object is *transcendent Being*, or 'God'; the most relative or contingent object is *matter*. In the same way, the absolute Subject is the *immanent Self*; the most relative or contingent subject is the *individual ego*. The corruption of the objective pole leads to materialism, and the corruption of the subjective to individualism. We reiterate, however, that the dictionary definitions of subject and object as given earlier will furnish the usage mainly followed in this work.

Thus for example, a material body whether acting or acted upon is for the purposes of this book never a subject. By way of illustration, an ocean wave demolishing a promontory of land is an *active agent* (cause)—hence *subject*—relative to the *passive object* it demolishes (effect). But this nonconscious wave in turn is the passive pawn of the wind, itself passive to high-altitude airstreams, which themselves are passive to hot and cold ocean currents, said to be governed in turn by the rotation of the earth, this rotation subjected to the dynamics of the solar system, which system is again but an element in a larger cosmic movement . . . .. Where then is the real subject? The only true subject here (Creator apart) can be a knower who registers these phenomena.[1]

To show the interdependence of what might be termed binary cognates, let us try to take in isolation the concept *altitude*. The height of Mt. Everest, for example, is variously given as 29,002 ft., 29,028 ft., and 29,141 ft. But conceived alone, what is a *height*? The very notion depends on the corollary of *lowness* or *depth* from which the height rises, the determining factor in this case being sea level. But this *level* factor in its turn is meaningless without recourse to the ideas of high and low, where 'sea level' is conventionally considered as being that measure *midway* between *mean high* tide and *low*. Otherwise, if measured from its base on the sea floor, then by total vertical length the world's 'highest' mountain is Hawaii's Mauna Kea (33,476 ft.), although from sea level its altitude reaches only 13,680 ft. All is relative.

Or to take an example from the domain of ideas: we can strive to imagine the concept of absolute nothingness. As Chuang-tse (d.*c*275 BC) puts it, 'At the beginning of the beginning, even nothing did not exist' (Ch. XII). If instead of symbolically this is to be understood literally (which it is not), then the Taoist sage has already erred on two scores, for a *beginning* is not a *nothing*, and two negatives make a positive. It is totally impossible to conceive of absolute nothingness, for the simple reason that a concept is already something, and presupposes a subject who conceives. If nothingness were absolute, it would indeed be a most tremendous something.

In regard to the term *evolution*, it is a perfectly valid concept as the corollary of *devolution*: a seed evolves into a tree; the tree devolves with time into humus. There are valid and invalid concepts: a square circle, like evolutionism, is an invalid concept.

On the indivisible nature of polarities, Mortimer J. Adler has this to say in his *Ten Philosophical Mistakes*:[2]

> The terms part and whole are indefinable. We cannot say what a

1. The absurdity of the notion of 'artificial intelligence' becomes apparent when one realizes that any 'thinking machine' is necessarily a created *object* which can only act on another object, in the way a hammer 'acts' on a nail. The unique subject in relation to any 'thinking machine' has to be the scientist who has constructed the apparatus. The animate (subject) can consciously shape the inanimate (object), but never vice versa. A machine has no ego.

2. New York, Macmillan, 1985, pp. 85–6.

part is without using the notion of whole, or what a whole is without using the notion of part, and so we cannot define either part or whole by itself. Nevertheless, we do so understand what parts and wholes are in relation to one another, that we cannot understand a part being greater than a whole or a whole less than a part.

Samuel Taylor Coleridge (1772–1834) has given a lucid definition of what he called the 'Law of Polarity':

> EVERY POWER IN NATURE AND IN SPIRIT *must evolve an opposite, as the sole means and condition of its manifestation*: AND ALL OPPOSITION IS A TENDENCY TO RE-UNION. This is the universal Law of Polarity or essential Dualism.[1]

Coming back to the pair subject-object, we have Schuon's metaphysical exposition on the Divine Substance:

> Since the Divine Substance, in virtue of one of its dimensions, 'willed' and 'had to' manifest the world with its multiplicity, it willed and had to manifest at the same time witnesses to this world and this multiplicity, otherwise the Universe would be like an unknown space filled with blind stones, and not a world perceived in accordance with a multitude of aspects. Where there are objects, there must also be subjects: creatures which are the witnesses of things belong indissolubly to creation. In its unfolding, the veil of *Mâyâ* strewed the void not only with things knowable, but also with beings capable of knowledge in varying degrees; the summit of these degrees, for our world at least, is man, and his sufficient reason is to see things as only an intelligence capable of objectivity, synthesis and transcendence can see them.
>
> . . . Where there are objects, there are subjects: that is why in Being there is an objective and passive pole, the principial *Materia*, *Prakriti*, and a subjective or active pole, the manifesting, determining and diversifying *Spiritus*, namely *Purusha*; and the same holds true, *mutatis mutandis*, at each level of the Universe.

1. *The Friend*, I.94n. Cited in Cutsinger (op. cit. in Preface), p. 63.

> . . . Having created the material world, God projected into it subjects capable of perceiving it, and finally he delegated man, who alone is capable of perceiving it totally, that is, in connection with its Cause or Substance; as a result, man is the measure of things, as all traditional doctrines attest. Man is situated, spatially speaking, between the 'infinitely big' and the 'infinitely small'—in Pascal's terminology—so that it is his subjectivity and not a quality of the objective world that creates the line of demarcation . . . .. Man is a point of junction between . . . the outward and the inward.[1]

Elsewhere in the same chapter, Schuon explains the sequence governing the process of manifestation:

> In the world, the objective element, *a priori virtual*, came before the subjective element *capable of actualizing it* by perception—Genesis testifies to this; whereas in the principial order, the subjective comes before the objective; this the world retraces in an inverse sense, precisely, since it is as it were a reflective surface [emphasis added].

The same idea finds expression in a quatrain by the great Persian Sufi poet Shabistarî (d. 1320):

> So the reflection of Absolute Being
> Can be viewed in this mirror of Not-Being,
> For non-existence, being opposite Reality,
> Instantly catches its reflection.[2]

Or as the *Pañcaviṁśa Brâhmaṇa* says (VII.8.1): 'There in the Waters Mitra-Varuṇa beheld-themselves-reflected.'[3]

The *time element* here is not measurable in years, but relatively intemporal, since God's operation is 'before' time has as yet coagu-

1. *Survey of Metaphysics and Esoterism*, Bloomington, IN, World Wisdom Books, 1986, the chapter 'Substance: Subject and Object' (this chapter first appeared in *Forme et Substance dans les Religions*, Paris, Dervy-Livres, 1975). 'The centre of all things is man, he is the middle point of heaven and earth' – Paracelsus: *Astronomia Magna* (Jolande Jacobi: *Selected Writings*, Princeton, 1988).

2. *The Secret Rose Garden of Sa'd-ud-Din Mahmud Shabistarî*, tr. Florence Lederer, Lahore, Ashraf Publication, n.d., p. 50.

3. Ananda K. Coomaraswamy, *The Transformation of Nature in Art*, Cambridge, MA, Harvard University Press, 1935, p. 210, n.96.

lated into the polarity before/after. 'Do not fondly imagine that God, when he created the heavens and the earth and all creatures, made one thing one day and another the next,' says Meister Eckhart (1260?–? 1327). 'Moses describes it thus it is true, nevertheless he knew better: he did so merely on account of those who are incapable of understanding or conceiving otherwise. All God did was: he willed and they were' (Evans, I.7).

> Verily His command, when He intendeth a thing, is that He saith unto it: Be! and it is (*Qur'ân*, XXXVI.82.)
>
> Nor before nor after was the process of God's outflowing over these waters (Dante [1265–1321], *Paradiso*, XXIX.20–21).
>
> He that liveth in eternity created all things at once [*creavit omnia simul*] (*Ecclesiasticus*, XVIII.1).

But why is it that, when evolutionism comes up, pseudo-science is allowed free reign to bypass all the basic coordinates of existence? It is because contemporary man is born with a mentality so exteriorized and centrifugal that he finds no problem in the premise that inanimate cosmic dust could 'evolve' itself into trillions of living organic cells ready to start the long journey up the ladder of transformism; that an *object*, physical entity, or *thing* 'out there', as near as is possible to the unrealizable concept of pure quantity, somehow manages to 'lift itself by its bootstraps' as it moves more and more towards the pole of quality (subject)—a possibility for which there is not the slightest shred of evidence in the entire phenomenal universe. To put it a different way: life can become lifeless, but not the inverse (except in the sense that the death of one life is the birth of another)—and we are not talking about miracles, these being outside of the subject treated here. Matter is not capricious: it is completely 'obedient' to the *laws governing* it. Insofar as creativity, freedom, and spontaneity exist, they belong to the qualitative pole of things—certainly not to the quantitative (terms that will be defined in Chapter 3).

Evolutionism would tell us—for this willy-nilly is the gist of its message—that the passive (*object*), inanimate, and purely material pole of existence, unaided by any external factor, gives birth to the active (*subject*) and animate pole, as if a purely feminine existence could create a masculine existence (or for that matter, a purely masculine a feminine), that the female pole (or for that matter, the male) is independent, autonomous, absolute.[1]

As for Pierre Teilhard de Chardin's contrivance of a primitive psychism attaching like an aura to the primal cosmic stuff, the first point to be taken into account is that Teilhardism is not Darwinism, and the second, that this theologian-paleontologist conceives of God (who logically has to be the ultimate in consciousness) as a being who evolves along with the creation—which notion contradicts the traditional conception of God as the Supreme Being, who by definition could not be Supreme if subjected to an ongoing process of development. And where and how for the rest did this primitive 'subject' or psychism originate in the unfolding of creation? The fact is, the cognizing subject paired to the cognized or objective universe must be at least commensurate in stature to the immensity of its counterpart, from the very beginning of time.[2]

Finally, in answer to the argument that there are objects which we as subjects can never know, that do not even cross our minds, the reply is that the Self alone as 'cognitive pleroma' knows All, within which framework our individual knowing is by remotion and analogy a limited or partial participation. As the *Praśna Upanishad* expresses it:

> Both what has been seen and what has not been seen, both what has been heard and what has not been heard, both what has been experienced and what has not been experienced, both the real (*sat*) and the unreal (*a-sat*)—he (the universal mind) sees all. He sees it, himself being all (IV.5, tr. Robert Ernest Hume).

To understand how contemporary man can feel comfortable with evolutionism, we have to go back in time, and even to his primaeval sources, to be able to determine how he has 'evolved' into the creature we know him for today.

1. The homosexual error is, among other things, that of isolating one pole of a binary cognate and treating it as an absolute, which does violence to the imperatives of the cosmic order.

2. For an exhaustive analysis of Teilhardism, see Wolfgang Smith, *Teilhardism and the New Religion*, Rockford, Illinois, Tan Books, 1988. See also the chapter on Teilhard in Malachi Martin, *The Jesuits*, New York, Linden Press/Simon and Schuster, 1987, and likewise the same in Philip Sherrard, see Suggested Reading.

2 *Creatio ex Nihilo*

> Have the disbelievers not seen that the heavens and the earth were a seamless whole? Then we parted them asunder. *Qur'ân*, XXI.30

A PECULIAR feature in Balinese temple architecture is the split gate (*candi*—or *tjandí*—*bentar*) which gives entrance to the first of two courts. (Fig. 1)

'Taken by itself, the *candi bentar* is surely one of the most disquieting structures in the architecture of the world,' writes Rudolf Mrázek.

> It is derived from the ancient Indian and Javanese *candi*, the sanctuary where the ashes of the important and powerful were kept and worshipped. The monument still retains the *candi* shape but, in contrast to both India and Java, in Bali the *candi* is split. All the ornaments lovingly carved on the front . . . of the *candi bentar* are mercilessly severed, cut through the middle of the smallest arabesque, and even through the middle of the grotesque face glaring from the top of the gate. The split gate is the passageway to the temple court, towards the gods, towards heaven. But there is a slash and something very pure and solemn seems to have been destroyed by it.[1]

Describing the same feature, Miguel Covarrubias writes that it

> was like the two halves of a solid tower cut clean through the middle, each half pushed apart to give access into the temple.

1. Werner Forman, Rudolf Mrázek, Bedřich Forman, *Bali – The Split Gate to Heaven*, London, Orbis Publishing, 1983, pp. 9–10.

> That the *tjandí bentar* represented the two halves of a unit was obvious; in most of them each side was elaborately carved, often with the design also cut in two . . .. Furthermore, the two inner sides were invariably left smooth, clean surfaces that shone by contrast with the elaborately carved rest of the temple. This we decided was an inviolable law until we found one *tjandí bentar* in Puru Bangkung, in Sukasada, North Bali, with its inner sides carved. This exception, however, is not important, given the anarchy that prevails in North Balinese temples, and since there is no rule in Bali without its exception.[1]

Covarrubias thus makes it amply clear that the accepted canon in Balinese temple construction prescribes that the inner side of the *tjandí bentar* be left imageless. The meaning of this architectural feature is inescapable: the split gate represents the sacrifice or diremption having to take place *in principio* in order that Essence become translated into Act; it signifies the primordial polarization that reverberates into the indefinity of polarizations constituting the heteronomous dualities of the created universe. Seen from the outside it is the 'Symplegades' motif of the 'Clashing Rocks', to which Ananda K. Coomaraswamy has devoted a study.[2] In itself it is the *barzakh* (isthmus, barrier) mentioned in the *Qur'ân* (LV.20 and elsewhere) separating the 'two seas' (the salt and the sweet, Lower and Higher Waters, Earth and Heaven) so that they 'encroach not one upon another.' From the inside it is the world of formless manifestation, hence 'left smooth.'

Within the Balinese temple compound a second monumental gate opens onto the inner court, and this portal, according to the same author, is 'identical in shape and design with the reunited halves of the *tjandí bentar*, but raised high above the ground on stone platforms, with a narrow entrance provided with wooden doors and reached by a flight of stone steps.'[3] The meaning here is axial, since this gate symbolizes access to higher states of reality, but

1. Miguel Covarrubias, *Island of Bali*, New York, Knopf, 1942, pp. 265–66.

2. See A.K. Coomaraswamy, *I: Selected Papers – Traditional Art and Symbolism*, ed. Roger Lipsey, Princeton, NJ, Princeton University Press, 1977.

3. Covarrubias, op. cit., *loc. cit.* This matter of the reunited halves is reminiscent, in a different context, of the Puerta Judiciaria (Gate of Judgment) giving entrance to the Alhambra in Grenada. On a stone arch is carved a key, and on an adjacent portal a hand.

the split gate is the one crucial to our study, which does not need to take into consideration integrally the manifold elements within the compound. What does have to be examined more fully is the inviolable nature of this 'Symplegades' or *barzakh* or 'eye of the needle', through which no 'plurality' or 'otherness' can gain entrance.

★

The nonmanifested, hence nondual, Subject—identical with the nonmanifested, hence nondual, Object—is the seamless Archetype of (and thus preceding) all the polarities fleshing out the interstices of manifested existence. It is therefore inconceivable that the creation or formation of the separative world could be attended by an evolutional process entirely independent of the law of polarities governing everything that has form, that there could in this one instance be an ultimately independent object without a correlative subject, when the principial Subject/Object is the very Source of all differentiations on a descending subject/object scale.

Returning to Schuon for further elucidation here:

> Spirit is Substance, matter is accident: that is to say that matter is but a contingent and transitory modality of the radiation of the Spirit which projects the worlds and the cycles while remaining transcendent and immutable. This radiation produces the polarization into subject and object: matter is the final point of the descent of the objective pole, sensorial consciousness being the corresponding subjective phenomenon. For the senses the object is matter, or let us say the perceptible physical domain; for the Intellect objective reality is the Spirit in all its forms. It is by it that we exist, and that we know; *were it not immanent in physical substances, these could not exist for one instant*. And in this Spirit, precisely, the subject-object opposition is resolved, it is

Legend has it that the day the hand finally grasps the key, Islam will again rule over Spain, all of which really signifies that on Judgment Day dualities will be reconciled. Curiously enough, at a later date a statue of the Virgin Mary has been placed above the key, which has its meaning in terms of the woman who shall crush the serpent's head (see *Gen.* III. 15) – the serpent here being the circumvolutions of our cycle's unfolding.

> resolved in Unity which is at once exclusive and inclusive, transcendent and immanent [emphasis added].[1]

In and above the Subject-Object hypostasis are all the polarities, dualities, complementaries, oppositions, and contradictories resolved. Nothing further can be said but that as qualities they *are*. 'I AM THAT I AM,' as God spake to Moses. The mystics, saints, and sages are not indulging in flights of hyperbole, but simply pure logic, when they attempt with a language restricted to polarities, to formulate nonpolarized truths. It is this that accounts for the negative or apophatic theology of Dionysius, and the nondual (*neti, neti*: neither this nor that) perspective of the Advaita Vedânta in Hinduism. As Johannes Scotus Erigena (815?–?877) puts it, 'God does not know *what* he is, because he is not any what.'[2] If a way of 'describing' God is to say that as Pure Being He is the Motionless Mover of all else, then what to say of the Supreme Principle, which is neither moved nor mover, motionless nor moving? According to Meister Eckhart, 'For God to see therein would cost him all his divine names and personal properties.'[3] Mahayana Buddhism uses the term 'Suchness' as a relatively nonlimiting way to express the Ineffable. 'Sameness is differentiation, differentiation is sameness,' says a Zen Formula. The Hesychast classic, *Unseen Warfare*, by Lorenzo Scupoli (1529–1610), tells us:

> Know that God is beyond all senses and sensory things, beyond all shape, colour, measure and place; is wholly without form and image and, while present in all things, is above all things; therefore He is beyond all imagining (I.xxvi).[4]

*

The state of gross manifestation and the subtle state, 'soul', or animating principle (*anima mundi*) which in traditional cosmology precedes the gross in hierarchy, both belong to the individual i.e.

1. *From the Divine to the Human – Survey of Metaphysics and Epistemology*, the chapter 'Consequences Flowing from the Mystery of Subjectivity', Bloomington, IN, World Wisdom Books, 1982.
2. In Coomaraswamy: 'Recollection, Indian and Platonic', *Supplement to the Journal of the American Oriental Society*, April–June, 1944, p. 13, n.37.
3. *Meister Eckhart*, tr. C. de B. Evans, London, Watkins, 1924, I, p. 38.
4. Tr. E. Kadloubovsky and G. E. H. Palmer, London, Faber & Faber, 1952.

separative or formal, and not universal plane of existence, and together could not exist without the interaction of the dualities and polarities of which they are composed. We could not even walk, for example, without there being a resistance at least equal to the thrust. Let alone talk, fight, love, eat, dream, compose, or anything else. In this so real-seeming 'sublunar' world down here, if the linchpin of dualities were once removed, there would be no more this world at all.

Everything positive in these correlations is principially contained on the formless plane of manifestation in the immutable Archetypes or Informing Essences, Divine Names, Powers, Exemplars, Energies, and Attributes, which lend their Qualities to outward existence without themselves being involved in their productions.

In the Unmanifested Itself, where all differentiation is excluded, God in the last analysis being, in Coomaraswamy's words, 'an inconnumerable and perfectly simple Identity', one can nevertheless posit the presence of a *principial* (as distinguished from the more relative *primordial*) polarity, which is what the Vedântists understand by the doctrine of Distinction without Difference (*bhedâbheda*). The Absolute has to be infinite, at the forefeit of its Absoluteness, and the Infinite has to be absolute, at the forfeit of its Infinity; and yet the Absolute is 'not' the Infinite, and the Infinite is 'not' the Absolute.

One might conceivably, then, envisage the Supreme Principle as a biunial pleroma, which could be represented by the hexagram—a single figure like a star, formed of two interlocked triangles inversely contraposed. Another image featuring essentially the same symbolism is the Tibetan Tantric *Yab-Yum* figuration of the Divinity locked in embrace with his Consort or *shakti*. In the case of the two triangles, the upward-pointing one could represent the Active Perfection of the Absolute Subject, and the downward-pointing one the Passive Perfection of the Infinite Object. The Infinite is in a way the adornment of the Absolute. The first determination of this plenary mutuality is found in the Hindu ternary *Sat-Chit-Ânanda* (Being-Knowledge-Bliss), which finds a variant formulation in the Christian Trinitarian doctrine. A further determination is that into Essence and Substance (the Hindu *Purusha* and *Prakriti*).

Now it is at this level, which still pertains to nonmanifested unity, that *separative* polarity originates, without however implying any duality in Essence and Substance. It comes about through a power contained in the Supreme Principle, a power designated by Hinduism as *Mâyâ*—this being the 'urge' outwards energized by the necessity in the very nature of the All-Possibility to manifest Itself, and which operates as a 'gravitational thrust' towards distinction, differentiation, individuation, and duality, thereby translating Essence into Act. Creation, thus viewed, is not unlike parturition on a cosmic scale. But from the standpoint of Heaven it is an awesome or 'dreadful' moment, since Self-Manifestation involves a sacrifice or 'slaying' or dismemberment of what has hitherto been united. Unity on a certain plane is shattered, Essence becomes Act, *nothing* becomes *thing* in the sense of *creatio ex nihilo*, the gate is split, and 'something very pure and solemn seems to have been destroyed'. The 'something' here is the Peace of what in Hinduism is called the 'Night of Brahmâ', when manifestation is withdrawn (the *pralāyâ*) at one degree or another into Principle; 'seems', because in fact the higher states of Reality are not affected by their modifications in space and time.

But it is none the less the 'sundering' of the heavens from the earth referred to in the *Qur'ân*, and the *dividing* of the waters referred to in *Genesis*, thus bringing into manifestation with its polarities and *dualities* the 'firmament', or what René Guénon terms the 'Empire of the Demiurge'. It is this aspect of creation that inspired Robert Herrick (1591–1674) to write:

> God hates the Duall Number; being known
> The lucklesse number of division:
> And when He blest each sev'rall Day, whereon
> He did His curious operation;
> 'Tis never read there (as the Fathers say)
> God blest His work done on the *second day*.

It is to this 'second day' that astronomers, unbeknownst to themselves and in that style or idiom peculiar to their profession, are alluding when they speak of the 'Big Bang'.

Jacob Boehme (1575–1624) comes closer to the truth and majesty of this moment in declaring that

> The Being of all beings is but one only Being, but in its generation it separates itself into two principles, viz. into light and darkness, into joy and sorrow, into evil and good, into love and anger, into fire and light, and out of these two eternal beginnings (or principles) into the third beginning, viz. into the

creation, to its own love-play and melody, according to the property of both eternal desires. (*Signatura Rerum*, XVI.8)

This 'love-play' is the same thing that Hinduism understands by the term *lîlâ*, meaning the motive of 'play' or 'sport' contained in the unfolding of creation, another incentive being that of God's desire for Self-Revelation. But anthropomorphically speaking, there is at the same time 'reluctance' on the part of the Celestial Powers to enter into manifestation, as this ineluctably brings into act the opposition of good and evil.

The descent into the world of forms is a rejoicing 'liberation', because the 'Waters' which have been pent up are now allowed to flow;[1] and it is an awesome moment, because the dualities contained in undifferentiated mode in the Nonmanifested are now unleashed. Thus on the one hand the *Qur'ân* lauds God for the fulfillment of possibilities: 'Glory be to Him Who created the pairs' (XXXVI.36); and on the other hand it admonishes believers to

> Say: I seek refuge in the Lord of the Daybreak
> From the evil of that which He created. (CXIII.1,2)

This last line is a stumbling block for those who would not have God permit—let alone be the author of—any evil: *'Si Deus est, unde Malum? Si non est, unde Bonum?'* On this matter, Schuon has repeatedly demonstrated in his writings that the Infinite by definition includes the possibility of manifestation, that manifestation means limitation, and limitation by its separativeness unavoidably includes suffering and evil. For within the All-Possibility, manifestation is inescapably one of the possibilities, and thus has its inevitable place in the world process.

God's 'repentance' of His creation (*Genesis* VI.6,7) is really a way of expressing His foreknowledge of the inexorability of evil and its consequences along with the rest. In other contexts, His Mercy outweighs His Wrath, as in *Psalm* CVI.45, where He 'repented according to the multitude of his mercies', and in *Jeremiah* XVIII.8: 'I will repent of the evil that I thought to do unto them'. To object that God, if Omnipotent and All-Merciful, would not have allowed evil to enter the world is quite simply without meaning. It is no more an

1. See Coomaraswamy: *The Ṛg Veda as Land-Náma-Bók*, London, Luzac, 1935.

insufficiency on God's part to be unable to create the pole *good* independently of the pole *evil* than it is an insufficiency on His part to be unable to create a square circle (the cosmic symbolism of the 'squaring of the circle' is not in question here), or a tree that is a bird. Whatever the evolutionists—wittingly or unwittingly—may claim about one pole of a duality being able to exist in the total and unqualified nonexistence of its corollary or counterpart, it is beyond the ingenuity of the Lord of Heaven and all His angels in concert to achieve this.[1] What God *has* given us with regard to good and evil is the discernment and volition to adhere to the good and eschew the evil. A proof of His perfection in existence, moreover, is that all the laws of matter function unerringly according to their properties, so that disequilibriums themselves arise from logical causes, even if these causes are not always identified or understood.

★

It should be clear from the considerations in the preceding pages that this matter of dualities has nothing to do with what rightly or wrongly is understood by the terms Manichaeism and Zoroastrianism, where the opposition between the Powers of Light and those of Darkness is thought to be irresolvable. Actually, as Guénon pointed out in his first published article, 'Le Démiurge',[2] Ormazd—the Wise Lord, and Ahriman—his twin brother the Evil Spirit, both proceed from the fundamental principle of all—Zrvana Akarana, in whom the two are merged at the end of time.

Polarities fall essentially into three categories:

(1) Those that are *reciprocal or complementary*, hence *neutral*, such as spatio/temporal dimensions, like right/left, in/out, original/reflection, before/after.

(2) Those that are *opposite but symmetrical*, like night/day, winter/summer, active/passive, positive/negative (in a functional and mathematical sense).

1. A symbolism of the 'pairs' in the Flood story is that manifestation was not to be consummated at the end of the cyclical period in question, but rather temporarily withdrawn 'into the Ark', and hence the polarities were maintained, but in a state of 'suspension' for an interval.

2. Paris, *La Gnose*, November, 1909; reprinted in *Études Traditionnelles*, June, 1951.

(3) Those that are *contradictory and dissymmetric*, like good/bad, beautiful/ugly, real/unreal, positive/negative (in a nihilistic sense).

When Eckhart declares that 'God is neither good nor true' (Evans, I.272), far from giving vent to audacious ellipsis, he is but announcing an incontrovertible spiritual truth. As was earlier said, higher realities elude the classifications accessible to polarized language. Which does not mean they are inexpressible, the vocabulary of symbolism and analogy constituting as it does a perfectly efficacious medium for the transmission of spiritual verities. But since these truths are not on the same plane with the polarized language, an intellectual effort on the part of the auditor is required to seize them, which exceeds the effort needed to grasp realities which are accessible on the same plane as the polarized language.

If God is 'good' according to what accepted parlance understands by the term, then one might think that by the law of dualities He must potentially, at least, contain the capacity to be evil, *quod absit*. This erroneous reasoning is resolved by understanding that dualities operate exclusively on the plane of cosmic manifestation or separation, given that disunion or division is the sole cause of their existence. Hence, if God is good, it cannot be according to our habitual comprehension of goodness, but rather according to what must be understood by the Good itself, which is Absolute, Uncreate, nonqualifiable Goodness. 'God is light, and in him is no darkness at all' (I *John*, I.5). Differentiation and otherness are uniquely the fact of remotion from Unity.

Since God is inconceivable apart from the quality of Pure Goodness, we have to conclude that evil is not really a quality at all, but rather the *absence* of a positive quality. As Guénon has explained in the same article, 'Le Démiurge', Good and Evil can only come into play in the outward domain of individuation and action, where change is involved. Since the supraformal realm is beyond all change, action is contained there *sub specie aeternitatis* in the simultaneity of its plenitude as Divine Act, in an *immutable* state where there is no longer any distinction between good and evil, so that 'the wolf shall dwell with the lamb' (*Isaiah*, XI.6), and which it would be completely erroneous to identify with immobility, since it is in reality what Eastern traditions understand by the term 'Non-acting Action'. The Cambridge Platonist, Peter Sterry (1613–72) puts it aptly:

> In this light of eternity alone is the Work of God seen aright, in the entire piece, in the whole design, from the beginning to the end. As all times appear in this light, less than a moment, a *point*, nothing, being as eternity in the undivided Unity of eternity; so are all the disorders of time, no more, not so much as a shadow in a dream that is past, but as the highest and sweetest Harmony in the undivided Unity of the eternal Harmony.[1]

How this can be is best understood in a static image like that of a cathedral, which is as a whole cosmos 'crystallized'. From the highest angel down to the gargoyles forming the gutter spouts, from the delicate turrets and traceries to the grandiose depiction of the Last Judgment on the West Portal, with angels and demons mixing among the elect and the damned, the entire ensemble inspires a marvellous sense of harmony and peace, where one essentially would not want a single stone other than it is.

1. Vivian de Sola Pinto, *Peter Sterry: Platonist and Puritan*, London, Cambridge University Press, 1934, pp. 175–76.

3 Contours of the Primordial Tradition

> Yonder world is in the likeness of this world, this world is the likeness of that.
> *Aitareya Brâhmaṇa*, VIII.2[1]

THE BURDEN of the demonstration thus far really comes down to saying that the level of existence characterized by manifestation, individuation, differentiation, and action, namely our world—the Sanskrit *saṁsâra* or Round of Existence—is *entirely* dependent on the heteronomous *interplay* of dualities (thus excluding any possibility, on our plane, of self-determination or autonomy), and of which the Prototype is the polarity *Subject/Object*. Other suitably equivalent terms for it are essence and substance (the Hindu *purusha/prakriti*), quality and quantity, the Scholastic form and matter, and the Hindu name-and-form (*nâma-rûpa*).

In man himself, the soul is *subject*, of which the body is *object*. One could also say that the link between *inward* consciousness, the subject, and *outward* body, the object, is *Life*. The total separation of subject from object here is immediate death. The subject or ego, however, is repossessed after dying, on another subject-object plane, whereas the previous object is dissolved, having terminated its function as receptacle for the subject, this being the only reason of its creation. In other words the subject as understood here primarily determines the object, and not vice versa. The subordinate dualities, moreover, cannot be separated without their mutual extinction.

From a panentheistic perspective (that Deity is both transcendent

1. Tr. A.K. Coomaraswamy.

and immanent) one could consider God as being the Subject whose 'body' is the object *world*, provided 'world' is taken to mean the substantial—in contradistinction to the essential—pole of existence in general. It is to this concept that Hermes alludes in his dictum:

> The universe is composed of a part that is material and a part that is incorporeal; and inasmuch as its body is made with soul in it, the universe is a living creature.[1]

Or as the Persian Sufi poet, 'Irâqî (d. 1289), expresses it:

> When in the glass of Beauty I behold,
> The Universe my image doth enfold.[2]

What is certain is that a cosmic dissociation between subject and object of necessity entails the instantaneous extinction of creation. In the words of the Welsh bard, Dafydd ap Gwilym (c. 1340–1400), 'A world without God is a world of nothing'.[3]

★

René Guénon in his book, *The Reign of Quantity and the Signs of the Times*,[4] has developed the doctrine of the interplay between subject and object as it relates to the modifications in the bipolarization between quality and quantity (an opposite but symmetrical duality).[5] Pure Quality in itself, being undiversified and formless, is situated above all manifestation, whereas Pure Quantity in itself can never be completely attained, since the realization of this would be entirely below manifestation and form, and hence indistinguishable from nothingness. But through the interpenetration and interaction of these two poles, the full play of diversity is effectuated. As Dante puts it: 'Form and matter, united and in purity, issued into being which had no flaw' (*Paradiso*, XXIX.22–3).

The reciprocity between the two poles quality/quantity can be represented by the *yin-yang* symbol, in Far Eastern cosmology, which

1. Lib. X.11; in Walter Scott, *Hermetica*, Oxford, Clarendon Press, 1924–36.
2. A.J. Arberry, *Sufism, An Account of the Mystics of Islam*, London, Allen & Unwin, 1950, p. 102.
3. *Blackfriars*, London, 1942, p. 476.
4. See Suggested Reading.
5. Obviously where two subjects are object to one another, the duality is not opposite but complementary.

is formed of a circle enclosing two interlocked helicoids, one white with a black dot in the center, and the other the inverse (Fig. 2). If we remove the dots, then we have another image of the biunial pleroma adumbrated in the preceding chapter under the form of the hexagram, and likewise the Tibetan *Yab-Yum* figure. In a *yin-yang* image without dots, the uniform white field would delineate the essential pole, or Pure (undifferentiated) Quality, whereas the uniform black field would delineate the substantial pole, or Quantity in the state of pure (undifferentiated) potentiality.

Figure 2: The eight Chinese Trigrams (schematic cosmic polarities) enclosing Yin-Yang symbol.

If now we add the dots, which serve in this representation to indicate the role of *Mâyâ*, we have an image of Essence translated into Act—a 'flow' suggested by the helicoidal form of the two fields. The black point (the 'serpent' in the Garden) in the white (qualitative) field functions as the minimal necessary separativity operated from the quantitative field to enable Quality to take on the distinction of forms. Thus the white progressively becomes 'black', and the black 'white'. In the unfolding of a cosmic cycle of existence, the movement can only commence from above downwards, descending from the qualitative towards the quantitative pole, from the higher towards the lower, from inward outwards, essence towards substance—and of which gravity is a figure.

Another similitude for this is found in the hourglass: here the upper chamber stands for the qualitative pole, and the sand for the indefinity of possibilities to be manifested in the course of a cycle, the lower chamber being the quantitative pole which makes their realization possible. The neck joining the two chambers functions as the *barzakh* or 'split gate'. As the flow of sand nears completion, the quantitative pole is overwhelmingly in the 'ascendent', to the seeming detriment of the opposite pole, which is nearing 'depletion'. This 'moment' or phase is manifested in the *yin-yang* pair as the black field determined only by the white dot. When the cycle is completed, the white dot 'disappears', the movement ceases, and 'time is changed into space', because, as Guénon says, where there is no more succession, there can only be simultaneity. But this 'moment' is actually that intemporal instant between two cycles, for it immediately engenders what is called the 'reversal of the poles', which is the beginning of a new cycle, no gradual return 'upwards' being possible. This is shown forth in the reversal of the hourglass: the sand starts descending again, imperceptibly at first, but progressively accelerating—the way we can experience time doing in our present world. It must be emphasized that the polarity in operation here belongs to the category of *opposite but symmetrical*, which means that the active or positive and qualitative pole will always in reality predominate over its passive (but not negative!) opposite, even when appearances suggest the contrary.

It is only the evolutionists who have had the imagination, or lack of it, to envisage the process in the inverse direction, from below upwards, outwards inwards, ascending from quantity towards quality, the higher evolving from the lower, the more proceeding from the less. And this, without even the need of a subjective pole to counterbalance the objective one and thus bring the motion into play in the first place.

*

It follows from the foregoing exposition that the subjective and qualitative pole will be very much in predominance at the beginning of a cosmic cycle, just as, by the law of inverse analogy, the objective and quantitative pole will—in appearance at least—have the ascendancy as the cycle approaches its end. The inference to be drawn from this is that a cycle will be characterized in its early stages by luminosity and transparency, making it propitious for the play of 'angelic' influences, and in its later stages, by increasing opacity and tenebrousness, making it propitious for the action of disordered and eventually subversive forces. In Hindu terminology it is the *guna* (one of three constituent tendencies in *prakriti*—undifferentiated substance) of *sattva* (the luminous and ascending aspect) that predominates in the early phases of a cycle; in the middle phases it is the *guna* of *rajas* (the expansive, fiery aspect) that predominates; while the latter phases are more particularly under the sway of the *guna tamas* (the descending, obscure aspect).

But cyclical succession must not be thought of as uniformly linear and uninterrupted; rather is it like a series of varying waves united only in the larger current that propels them onward. Also, it is the indispensable combination of the three *gunas* functioning in concert that maintains the fundamental equilibrium of polarities.

★

Scriptures both Western and Eastern are in agreement that through Utterance God brings forth the universe: it is the resonance of the primordial Vibration upon the stillness of the Void (Waters), by which the worlds are formed. The Word spoken is the supreme Sound; it is the Divinity's own Name and Form, the Logos, Wisdom.

This is the Spirit or 'breath of life' He breathes into the 'dust' out of which Adam is formed as a 'living soul', the 'dust' being primordial matter—*materia prima*—before its differentiation into the outward elements. This first human being transcends individuality as such; he is what various traditions refer to as Universal Man, being created 'in the image of God', or as the 'vicegerent' of God—to use the Qur'ânic term, and who accordingly is a microcosmic abridgement of the *entire* universe.

Through a further exteriorization he is polarized into man-and-woman and thus is on the threshold of individuation. The first man-and-woman may remain in that threshold state as long as 'they' obey God's injunction to abstain from the fruit of the Tree of the Knowledge of Good and Evil.

Now speaking anthropomorphically, it is not to be supposed that God, when He admonished the first human couple against tasting the fruit of that tree, entertained any illusion about the outcome, no more than a parent, placing a child before a Christmas tree, with the injunction not to touch, could imagine that it would be obeyed.

Moreover, it is only through this act of 'eating' that the dualities of discursive knowledge can be brought into play and thus allow God to 'enjoy' the vision of Himself in external modality, which is the whole basis for outward existence.

But it was incumbent on primordial Man—vicegerent, the image of God, and possessed as he was of the state of centrality as also of free will—to bear the immediate responsibility for the Fall, or the plunge into the realm of dualities. In the contrary case, had there been no warning, Adam and Eve could have considered themselves the 'victims' of a 'caprice' on God's part and abjured any obligation for repentance and restitution. In other words, they could have claimed that God alone as the Author of Good and Evil was to blame for the ills inevitably ensuing in the world that He—not they—had put into motion.

If these words convey a familiar ring, it is because they in fact bespeak the spiritual or moral posture into which modern man has maneuvered himself: science has convinced him that we are but ciphers in a space extending billions of light years throughout a universe that may or may not include a God somewhere in the scheme of things. Therefore to modern man, the reality of the situation is that—Creator or no—we are by all evidence the pawn of an evolutional process (or whatever you may wish to call it), born of chaos, that leaves us little option but to exploit the 'obvious' status quo in good conscience or otherwise, and master this environing chaos according to *our* ingenuity.

Scripture may well say that *God* created Adam of dust—the *Qur'ân* in this context includes Jesus (III.59); but on the evidence of evolution it is the *dust* itself which produced both Adam and Christ (or at least produced *homo sapiens*, in deference to those who have scruples against sacred history). Only, in bypassing the Creator, the evolutionists, in order to launch their theory, still must needs resort to a *deus ex machina*, an 'entity' which they offer us under the name of Transformism. And even so, they are not really presenting anything new in the way of a positive contribution, since substantially all they have come up with is a version of Genesis inverted, or what amounts to the same thing, the Gospel of St. John inverted to read: 'And the flesh was made Word', which to those not of their persuasion can be none other than blasphemy.

★

A universal symbol featured in all traditions is that of the Tree at the Center.[1] This is the axis running through the worlds that links them to their uncreate Source. Although sometimes represented by two trees, as with the Tree of Life and the Tree of the Knowledge of Good and Evil in *Genesis*, these two are really the selfsame tree, depending on its being viewed under the aspect of unity, or the aspect of diversity. The image of the Christmas tree in the preceding section was used advisedly, since whether people know it or not, their ritual of decking out an evergreen in the parlor at the time of the winter solstice is originally in token of the World Tree or *Axis Mundi*.[2]

Adam is created a living *soul*. In Plato's (*c*428–348 BC) words, 'Our original nature was by no means the same as it is now . . .. For 'man-woman' ('ανδρόγυνον) was then a unity in form no less than name.'[3] 'And he called *their* name Adam' (*Gen.* V.2). But at the point of the Fall, Adam-and-Eve stand before the Tree, 'themselves' on the threshold of individuation, and are willingly captivated by a vision which is polarizing from unitive to separative. This happens in the 'cool of the day', namely, when the Supernal Sun or fullness of the Beatific Vision is receding. They now perceive their nakedness, that is, their 'otherness' as individuals. God gives them 'coats of skins', meaning that they have crossed the threshold of the 'split gate' and entered fully into the state of corporeal manifestation, with physical bodies.

Something of the luminosity of these bodies, though, can be deduced from Boehme's words concerning the flesh and blood in question:

> This matter was real, but in the same way that shadows are latent in the light, which does not alter their inability to manifest themselves in the presence of this light; but when the light is

1. See for example *The Tree of Life, Symbol of the Centre*, by Roger Cook, London, Thames and Hudson, 1974.

2. This tree, moreover, neccessarily had to be at the North Pole, as it is not conceivable metaphysically that the Terrestrial Paradise could be situated elsewhere than on the world axis. See in this respect William F. Warren, *Paradise Found, the Cradle of the Human Race at the North Pole, a Study of the Prehistoric World* (Boston, 1885); and B.G. Tilak, *The Arctic Home of the Vedas* (Poonah City, 1925).

3. *Symposium*, 189 E, tr. A.K. Coomaraswamy.

diminished the shadows appear (*Mysterium Magnum*, XVI.6 —from the French of Nicolas Berdiaeff).

It is not the purpose of this dissertation to enter into a theological exposition; and what has been written here is purely to impress on the reader from the evidence of Scripture the overwhelmingly 'subjective' nature of our 'first parents', at the same time demonstrating that their *subjectivity* was totally the opposite of the 'subjectivity' popularly ascribed to early man, by what can only be called the anthropological bias of our times. If the beatific vision or undivided awareness of the presence of the Supreme Center was no longer retained integrally, these people of the Golden Age were none the less still plunged in the reverberations of it.

What characterizes the humanity of the Primordial Tradition is the existential awareness it continued to have of God in His unicity, and of phenomena as the manifestation of His attributes. Its vision remained essentially *nonseparative*, being the *predisposition* to see things in their theomorphic transparency. For early mankind as *subject*, the complementary *object* was less the world than God, along with His traces in creation. Matter in itself at this period hardly existed as an object 'out there'; it was as yet so unitive and qualitative as to be quasi-translucent and scarcely congealed. To return to the *yin-yang* symbol, matter in a concrete sense was little more quantitatively ponderous than the black spot's presence on the white field. We say these things not by way of intuition but simply by reversing the world picture as it is now.

And yet throughout all the periods of a cycle's passage the subjective or inner pole, of which the objective or outward one is the counterpart, cannot by definition be less real than its objective counterpart, which remains wholly determined by the inward pole. There always has to be a proper adequation between subject and object. It is not like the pair real-unreal, where one pole only is 'positive', and the more it is so, the more 'negative' the other. If this adequacy were lacking, no conscious creature upon contact with the material order could meet with other than chaos and speedy death.

Early man, of whom we still have the traces, if not the understanding, was necessarily a being perfectly adapted to his environment, which was for him Center and home, and not an alien space inhabited by hostile beasts, forces, and spirits that had to be propitiated through rituals and ceremonies born of ignorance and superstition. This is not to say he was unaware of the subtle realm, but simply that it all formed part of the universal harmony.

Concerning those protocreatures whose vestiges and fragments are the subject-matter of paleontology: whatever one may theorize about them, it is certain that they likewise had to be perfectly adapted to the exigencies of their surroundings. This means they had fins, fur, wings, eyes, teeth, a digestive system and so forth completely developed and suited to the functions required. The same goes by analogy for the vegetable kingdom. The only door open to a life-form ill-adapted to its ambience is extinction, examples of which are in ample evidence with us right here and now. As for a diamond, a cedar, or a lion, how can one imagine there being a need or initiative to 'mutate' into something 'higher'? Do nightingales sing and eagles soar better today than did their first progenitors?

Given what we know about ecology in our present world, it is only logical to presume that the most ancient creatures whose traces remain were in their day all part of our perfectly coherent 'ecosystem'—otherwise how explain the fragile balance in the biosphere that has been preserved and maintained, despite major environmental permutations, from their time up to ours? And what can we know of those forms of life that vanished in a universe less solidified than it subsequently has become, not to mention other forms that disappeared with previous cyclical cataclysms?

*

There is more that needs saying about these early creatures. It was pointed out in the first chapter how the sequence of manifestation reflects inversely the principial order of hierarchy, that is, the peripheral creation precedes the central, as various traditions testify. It is in the way that an artist conceives an idea before its execution. An architect first presents builders with the plan of a palace, which they then erect from brute stones, after which the edifice is provisioned and the courtiers assembled, with finally the King and Queen themselves arriving and being enthroned therein. So likewise does God finally place the crown jewel of His creation—Man—in the Garden already prepared with its panoply of manifestation to receive him.

But from the moment the Spirit first breathes on the waters, the entire creative process is uniquely by way of descent, and each life-form that subsequently follows comes as a fresh and spontaneous creation, being in every case that which it will remain for the rest of its terrestrial term—a *mutationless* shadow of the particular *immutable* archetype from which it derives (*Gen.* II.4,5). The *soul* or spirit in any creature (the immanent Self 'trapped' in its productions) may *aspire* upwards, in the sense of St. Paul's 'the whole creation groaneth' for salvation, but the *volitionless body*-form invested on each phylum of creature is fixed in that form forevermore—no evolving, mutating, transforming—the physical typology in every case being neither less nor more than it is meant to be, but adapted with a finality perfectly suited to the purpose intended for it. The chicken, in other words, comes both 'before' and 'after' the egg: first as archetype, afterwards as creature. The flow is always from the Center to the periphery: the sun energizes the plant, the heart energizes the body, the Supernal Sun energizes the physical one, the inner heart the outward.[1] In the words of St. Thomas Aquinas (1225?–1274), 'No creature can attain to a higher grade of nature without ceasing to exist' (*Summa Theol.* I.63.3), which means that it cannot 'evolve', physically but must 'transmigrate' spiritually to reach a higher form.

It was moreover stated in Chapter I that the whole creative process is in some way extra-temporal. Only when primordial man has fully entered into the spatio-temporal dimension is the formative act with its exteriorization of polarities completed (figuratively, the 'Sab-

1. This principle holds throughout organic life. Thus, what is known in homoeopathic medicine as Hering's Law proclaims that '*cure proceeds from above downward, from within outward, from the most important organs to least important organs, and in the inverse order of appearance of symptoms*' (George Vithoulkas, *The Science of Homeopathy*, New York, 1980, p. 231). The native American shaman observes the same healing sequence, starting with the subtle space just above the crown of the head, then on down through the body and limbs, and ending with the feet. Morbific poisons are brushed out with an eagle feather consecrated for the purpose, and sometimes with the addition of animal skins containing certain healing properties.

The celebrated American homoeopathic physician, Dr. James Tyler Kent (1849–1916), says in *Lecture* IV of his *Lectures on Homoeopathic Philosophy* (New Delhi, 1977): 'Every cell in man has its representative in the innermost, the middle and the outermost; there is no cell in man that does not have its will and understanding, its soul stuff or limbus or simple substance, and its material substance.

'Disease must flow in accordance with this order, because there is no inward flow. Man is protected against things flowing in from the outward toward the centre.' For further on this, see my article 'The Alchemy in Homoeopathy', *Studies in Comparative Religion*, Winter – Spring 1984.

bath'), after which the creative function, according to traditional theology, continues uniquely in the sustention to the extent possible of its myriad productions.

★

Credulity is indigenous to human nature. Schuon calls this frailty 'the human incapacity to exert the intelligence on all planes at once'. Man's *naïveté* begins where his competence leaves off. If people of earlier centuries imagined absurd creatures inhabiting unknown continents, men of later times have no less absurd ideas about these earlier men. As someone said, it takes greater faith to believe in evolution than it does to believe in God. Not to mention the proclivity of modern man—and accredited astronomy is not immune here—to conjure up in his imagination crazier beings inhabiting outer space than ancient man could possibly have conceived of.[1] Moreover, as Guénon has pointed out in *The Reign of Quantity*, the world did not always present the same topography it now does, having become more solid and opaque over millennia, apart from appreciable modifications in land masses, as is well documented today.

The same author also challenges contemporary man's claim to set the limits to the modalities of existence possible in earlier periods, especially given the closer interpenetration that formerly obtained between the subtle and the physical domains.

When Genesis tells us that God had Adam name all living creatures, the logical inference from this has to be that primordial man knew the inner properties of things, since a name, to be meaningful, has to denote the essential characteristic of the object named. It is here, then, that we have the archetypal pattern of the subjective pole 'naming' and thus 'legitimizing' the objective pole of existence. This means that early humanity knew the properties not only of plants, but of minerals and metals, and by extension the visible movement of the heavens, and of how the Zodiacal configurations determine the multifaceted interplay of sublunar life.

1. The credulous mentality in question is exemplified in projects afoot to beam radio signals and position microwave antennae towards other galaxies in the expectation that a similar effort may be in operation 'simultaneously' by creatures from these regions trying to 'reach out and touch someone'. Despite his deeply ingrained materialism, contemporary man instinctively senses the need for a subjective pole existing throughout the universe.

What for us are extraordinary feats of navigation on the part of the ancient Polynesians were for them but the natural way of covering the Pacific—given the coordinates of cosmic properties perceptible and exploitable within the framework of their time.

And what for us moderns are the blundering inaccuracies and deficiencies in the maps of yore as compared with modern cartographical precision, this 'inaccuracy' (even supposing it had no symbolic content) still did not deter the grandiose military campaigns, expeditions, and explorations of figures like Alexander the Great, Julius Caesar, Ferdinand Magellan, or Simon Bolivar. Napoleon Bonaparte himself, when it came to communications, effectively integrated the major portion of his vast military exploits without the assistance of Samuel Morse—something that would be unthinkable nowadays. The point being made is that a perfect adequation or working equilibrium between subject and object is never lacking at any time in world history; man by inalienable birthright is an essentially sophisticated creature. His problem is in not being sophisticated enough to realize that sophistication has multiform possibilities. It is altogether another matter that some modalities between subject and object may be more propitious or felicitous than others: one cannot, for example, imagine Tartar horsemen taking kindly to driving armored tanks—even though, given the sophistication just mentioned, anything is possible, including a sophistication that goes counter to man's final ends. It must not be overlooked, moreover, that we moderns belong to a society characterized by Coomaraswamy (in his essay 'What is Civilization?') as 'a world of "Epigoni", inheritors, rather than creators of any positive goods'.

*

In a chapter on Cain and Abel in his *Reign of Quantity*, Guénon defines Cain as the prototype of the agricultural, or *sedentary*, way of life, with Abel representing the pastoral, or *nomadic*, way. The first is fixed more particularly in terms of time (as in the Hindu preoccupation, for example, with cosmic cycles), and the second in terms of space (with peoples like the desert Bedouins and seafaring Vikings). Where compensating factors are lacking, the sedentary tends towards a certain solidification under the principle of compression, represented by time (we speak, for instance, of 'the pressure of time'), whereas the nomad tends towards a certain dispersion under the principle of expansion, represented by space (we speak of 'space expanding', which connotes a centrifugal movement away from the Center in the direction 'out there'). But normally these two ways share in a spatio-temporal complementarism that helps maintain the balance: thus the 'temporal' and 'concrete' sedentary exploits the spatial and visual plastic arts, such as architecture and painting, while the 'spatial' and 'abstract' nomad excels in the temporal and rhythmic phonetic arts, such as music (with song and dance) and poetry. These ways, moreover, are not closed systems, it being a matter of accent and tendencies. The cosmic tendency, however, is that of time encroaching *progress*ively upon space as the cycle advances. There is no 'progress' with the nomads, for whom the outward is essentially a dimension of the inward; that is why they generally are regarded as 'backward'. They typically, for example, make little or no use of the wheel, although this idiosyncrasy is not limited to nomads.

The ancient Polynesians referred to above were, like the Vikings, nomads of the sea; unlike the Vikings, though, their Pacific migrations over a period estimated, despite the lack of written records, to be some 3000 years, took place in splendid isolation from the peoples of the land civilizations, with whom they became independent of all contact, as if inhabiting another world.[1] During this time they fully populated the sparse volcanic islands and atolls in a watery triangle covering 12 million square miles—a solitary and intemporal space of sky, ocean, and island greenery, where the spiritual center of the gods and ancestors was a region called Hawaiki.

Representative are the Tahitians, a spirited people whose nomadic arts featured dance, drama, and especially oratory: the orally transmitted legends, epics, chants, and sagas of the Polynesians are among the finest to be found in the literature of the world. And presiding above and throughout this the gods, beginning with Taaroa, ruled supreme, for the Polynesian's—hence Tahitian's—cosmic vision was inwardly oriented.

Thus these people were still living—albeit passively—in the reverberations of a far earlier age, when intruded precipitantly on their shores in the mid-eighteenth century the vanguard of a sedentary civilization from the opposite side of the globe, and one as outwardly

1. Documentation for the material that follows is drawn largely from *Tahiti – A Paradise Lost*, by David Howarth, London, Harvill Press, 1983; and from *The Rape of Tahiti*, by Edward Dodd, New York, Dodd, Mead & Company, 1983.

oriented as the world at that cyclical moment could possibly produce. With it came those elements most grossly representative of its culture: iron, of the Iron Age (the Hindu *Kali-Yuga*), cannon, venereal disease, liquor, commerce, and creeds that were, to put it mildly, ill-suited to the spiritual way of the Polynesian pantheon. For once Cain was opting to be his brother's keeper, precisely when Abel least sought or needed it.

It was in 1797 that the good ship *Duff*, hailing from Portsmouth, England, disembarked eighteen members of the London Missionary Society in Matavai Bay, Tahiti, this being just ten years after Captain William Bligh had set out in the HMS *Bounty* on his ill-fated voyage to the island. The Society in question had been founded two years previously, its members consisting of Methodists, Baptists, Calvinists, Wesleyans, Presbyterians, and other dissenters from the Established Anglican Church, and composed for the most part of rugged artisans from London's poorest parishes.

They needed 'savages' to convert, and this was exactly the windfall the discovery of Tahiti brought them: fallen man living by their criterion in the cesspools of *sin*. That these happy, innocent, eloquent, skilled, and highly organized people still bore the traces of paradise, could nevertheless not go unremarked by the missionaries, who to keep their conditioned bearings had to repeat by rote the tidings of gloom and doom which had been deeply inculcated into them before departure from England. These men of the cloth told the natives that they, the missionaries, had been sent by God to bring the savages out of the darkness into light. But this message did not penetrate, since for the Polynesians *po* or darkness is the realm of the Sacred, whereas *ao* or light is the region of everyday life. When the Tahitians learned about the Fall of Adam with the consequence of his being driven out of Paradise, it only made them the more grateful that Taaroa had never cast *them* adrift from the Presence of the Sacred.

They were taught the Ten Commandments, but as David Howarth shows, the Decalogue in its turn fell on unbelieving ears. The Tahitians had never given any of their gods precedence over the Supreme Being. Their graven figures were too highly stylized and abstract to be 'images' in the literal sense, and they served as spiritual presences, not objects of worship. Certainly they never took the name of their Divinity in vain. It was futile to tell them to observe a Sabbath they knew nothing about, or to rest from labors which they did not regard as toilsome. They honored not only their fathers and mothers, but as a tribal people, all their kith and kin. Practically unknown to them was murder, which term for them would not be applicable to ritually controlled occasions of infanticide and sacrifice. What we call adultery was in their case circumscribed by tribal laws and taboos (*tapu*).

Having no concept of private property, they had little concept of stealing. It is true that they ran off with Captain James Cook's quadrant, thus imperiling the Royal Society's project to have a South Pacific sighting of Venus transiting the Sun on 3 June 1769. But the indispensable instrument was returned in time, the whole lark having been a feat of derring-do on the part of the natives to see who was valorous enough to penetrate an armed fort at night and make off with the mysterious bulky instrument. Moreover, only Europeans, never Tahitians, were treated in such wise.

This leaves us with false witness, which it was not in the Tahitian character to bear, and with covetousness—hardly a problem, given their social structure. These island people, to say the truth, were frankly mystified by the presence of the missionaries. Was a shortage of women in their homelands possible, since the ships which kept arriving were mostly filled with men? But the native women were not sought after. Or was their soil too sparse to grow the crops needed, as the departing boats were always laden with the produce of the island? Whatever the answer, it was not to be found in the bland church services, preaching, and solemn hymn-singing.

Only with time, suffering, and confusion would it dawn on the more observant of the Tahitians that religion was all part of a package with these outwardly oriented Westerners to promote their deity—Materialism—in those corners of the world that still lacked it. And certainly be it said on behalf of the missionaries that, in their self-delusion, the last suspicion to cross their minds would be the idea they could in any way represent what their own Scripture refers to as 'whited sepulchres'.

The observation to be drawn from the preceding passages is that inwardly oriented but passive nomads are no match at our time in the cycle for outwardly oriented and aggressive sedentaries. In earlier more normal periods things were otherwise. Belligerent nomadic peoples like the Vikings and the Mongols often brought fresh infusions of vitality into the sedentary and sometimes decadent civilizations which they conquered. But the increasing compaction and density of the material environment in recent centuries, accompanied by the quantitative expansion of populations, has deprived the nomad of the solitude and spatial conditions requisite for his spiritual outlook, with its consequent way of life—a way romanticized by a world nostalgic for that from which it has irrevocably cut itself off.

*

Nowhere has this clash of alien cultures been more apparent than in the brutal and disastrous confrontations on the soil of the two Americas, in the largely blind encounters between the sedentary settlers from Europe and the indigenous nomads. The Indians were masters of the martial arts, but of what avail is any qualitative art when faced with the massive fire power of quantitative weaponry? In the name of Manifest Destiny, the adventurous English, French, and Spanish sedentaries (accompanied, of course, by their missionaries to put a good face on this Destiny) went charging full speed into the outward wilderness, resolved to eliminate anything and everything that might prove an obstacle, or merely a target, whether virgin space, buffalo, or Indians (including those whose civilizations were of a sedentary character).

The traditional Native American inhabited the center of a spherical expanse delineated by the six directions of space, at the invisible Center of which dwells the Great Spirit. Everything came to him from this space, his dreams, his visions, his nourishment, his prophecies, his exploits, dances, and songs. Everything in this space was sacred, the animals, the birds, the plants, the rocks and land and water. The Indian is a lingering remnant from the Primordial Tradition, a being sadly unsuited to the material exigencies of the *Kali-Yuga*. Until now he experienced little essential break between subject and object, inner and outer. And there are still a representative number living who have retained somewhat of this perspective.

Anyone who has witnessed the rites and ceremonies of the Plains Indians will have remarked on the very relative importance the presence of time has to the participants, and how the spatial element for them is the essential, and furthermore how this space is conceived as being a continuum between the subjective world and the objective.

It is a perspective, as was said, that lingers in one way or another with Indians in all tribes who have maintained a contact with their heritage. Some years ago I met a group of Iroquois aged six to eighty from Ontario, invited to exhibit their crafts and dances at the Palais de Beaulieu in Lausanne under the auspices of a Swiss-Canadian cultural exchange. A lady who wished to meet these people requested that I accompany her to the exposition in the hopes of approaching their chief. They said they had no chief, but introduced us to the leader of the group, a big man of serious mien in his forties named Johnny Two Rivers, who drew up chairs and asked what he could do for us.

'Well,' said my companion, 'we would be interested to hear what you could tell us about your Indian relig . . .'.

'Religion?' he interrupted before she could finish, 'Religion—it's a crutch for people who need it. A crutch is better than nothing, but people who can walk on their two feet spiritually don't need a crutch. If you see the Great Spirit in everything, if it acts in everything you do, then you don't need this crutch!'

Now obviously the Iroquois have their religion; but this was an elliptical way—somewhat in the manner of a Zen koan—of stressing the primordial and cosmic or 'shamanistic' vision essential to the Indian perspective. He made this remark, moreover, in the full knowledge that the majority of Indians in his group were devout adherents of Roman Catholicism (even if they had just consecrated a new church with a 'big buffalo dance!'). Religion in fact is a generic term covering just about everything having to do near or far with the bonding of man with God. It all depends on what you mean when you use the word.

If we say for example that Christianity was the faith running like a sap in the veins of Mediaeval Europe, we are certainly not speaking about anything commonly understood as Christianity today. What was then the Christian civilization or the Christian *world* became, with the *secularization* introduced by the Renaissance, the Christian *religion*—a worship relegated to monasteries and churches as being, precisely, *otherworldly*. And what from that time on remained the thing known as the Christian religion has finally become, with our sacramenting of modern science—let us make no bones about it, it is just that (along with the exaltation of consumerism, relativism, psychologism)—the sordid and suicidal secularization of faith itself. Thus the objective world or pole has now virtually become the center, and the subjective one the periphery.

Later in the conversation with our Iroquois interlocutor, I asked whether he had close contacts with that sector of his tribe located across the border in New York.

'What's the United States?' he rejoined; 'what's Canada? It's just a line drawn with a pencil across a piece of paper!'

Again, he was not 'putting it on' to be pedantic or lofty, but was simply expressing a nomadic vision of space anterior to the encroachment on it of time, division, and possession. An older Indian with silver braids later told us that his people were originally hunters. Then the white man came and took away their forests, and so they turned to farming. Then the white man took away their land, and so now they build skyscrapers (Iroquois and other Indians being employed for this because of their freedom from vertigo).

*

A remarkable witness of a sedentary people maintaining an intimate identity between inward subject and outward object, and keeping their traditions essentially intact at least up through the first third of our century, amidst the criss-crossing of world trade routes and rivalry, is that of the Balinese. One could almost say that they have an 'obsession' with time only matched by their 'horror' of space. And yet this perspective, which offhand appears downright topsy-turvy to the normal nomadic outlook, is probably a crucial factor in the die-hard longevity of their civilization.

The Balinese religion can alone be situated properly in terms of a total life-style derived from a felicitous amalgam of animism, Hinduism, and Buddhism, with a modified form of Shaivite Brahmanism in the predominant role. More exact than saying the Balinese 'practice a religion' is to state that their whole social structure literally exudes a perfume of the supranatural. Taking together public and private shrines, for example, it is estimated that upwards of 40,000 temples are spread upon this tiny island of some 2000 square miles, to serve a population numbering little over a million souls.

The ancient Indonesians or 'Proto-Malayans', whom anthropologists relate to the primitive Polynesians and Melanesians, probably came to Bali from Southern China around 2000 BC. Their direct descendents are called the Bali Aga, who mostly inhabit the more inaccessible mountain villages and who still strongly retain their ancestral animism, minus the nomadism.

When Islam conquered Java in 1478 AD, this brought about the collapse of the ancient Hindu Majâpahit dynasty, which forthwith transplanted its whole spiritual and cultural heritage—the priesthood with its Hindu and Buddhist elements, the court and aristocracy, the sacred arts and architecture, music and dances—onto Balinese soil, absorbing the Bali Aga and fusing its primal animism into the fibers of the assimilating genius. This happy blend of elements apportioned itself among the castes, high and low, in such manner that each sector of society received the *dharma* or *way* for which it was most qualified.

As the majority of people in any sedentary *Kali-Yuga* civilization that has not degenerated into a predominantly bourgeois mold are peasants, a religion that insufficiently includes their needs, interests, and aptitudes will eventually find itself in conflict with what becomes the 'masses'. The unspoiled peasantry of Bali was keenly attuned to the inner world of animic influences, and the highly Brahmanical and intellectual priesthood functioned in its turn to aid the agrarians in keeping these influences properly channelled and neutralized from deleterious magic and contaminating elements.

Animism relates to the spiritual practices resulting from an awareness of the subtle (animic) forces underlying the outward forms of matter and nature in general. What is popularly taken by us for backwardness and superstition on the part of traditional peoples in every part of the globe holding these views is in fact not that at all, since it is hardly a mark of ignorance to be conscious of psychic influences informing what Westerners foremost in their materialistic blindness regard as purely inert 'matter'. Were modern man in fact less totally ignorant himself in this domain, his enthusiasm to wrest every last secret out of physical substances and exploit them for his technological and utilitarian ends would be tempered by the understanding that in so doing he exposes himself passively to becoming the victim, exploited—one would like to say 'possessed'—by entities of whose existence he has not the slightest suspicion. And yet anyone in our present civilization whose discernment is not completely atrophied must by force of evidence see that the humanity of 'our time' is contaminated with an inner (psychic) pollution that spills over by exponential shockwaves into all aspects of outward society. These things are not fortuitous or accidental. We live in a 'junk' world, 'evolution' notwithstanding.

An objection might be raised that if material substances are only the corporeal projections or outer shell of subtle entities, then evolutionists are entitled to claim these psychic properties as being the *subjective* pole pre-existing in the primal *objective* dust. But this argument is threefold groundless: (1) evolutionists in fact have (or believe they have) no truck with psychic entities; (2) the very presence of a subjective pole, as pointed out in the first chapter of this book, would void their transformist hypothesis; (3) the subtle prototypes of outward forms are exactly that: the informing principle on the individual plane below the formless Archetypes that gives its stamp to the outward thing, so that, precisely, iron is iron, and crystal crystal.

It is the subtle and gross forms together that make up the integral object, which is without a *separate* or independent subject as such. Iron is assigned by our analytical science an 'atomic weight' which it is 'condemned' to bear in perpetuity, and it is not the same weight as that given to copper or any other element. The subtle principle in iron cannot 'mutate' across inorganic possibilities, and all the less so from inorganic to organic. One might at this point refer to the supposed metallic transmutations of the ancient alchemists; but whatever they themselves understood by this, their position was unqualified: there is no transmutation without a given substance, such as lead, first being integrally reduced to its original *materia prima*, from which *formless* state only can a new or 'glorified' form, such as gold, be produced. If for 'lead' one reads 'human soul', then at least their spiritual message is clear. The part can go back to the Whole from which it originated, but it cannot become another part or substance without first passing through 'extinction'—in keeping with the dictum of Aquinas given earlier in this chapter.

To return to the Balinese, they have an inherited tendency to be keenly aware of both the angelic and the dark powers emanating from the invisible world; and the multitude of rituals, sacrifices, dances, and offerings they impassibly perform is in view of gaining the spiritual favor of their patron gods while at the same time and with the same impassivity placating the resident demons by tossing them from these offerings the material bone insatiably coveted.

In other words, the people of this island know they are in the *Kali-Yuga*, and they avail themselves of this knowledge to neutralize the *moment* through a highly elaborate chronology, regulated mainly by two calendars—the *saka* for the solar-lunar year, and the *wuku* for the year based on weeks—a temporal system which only the priests fully understand, but which is exploited in the ordering of all aspects of their life: cultivation, irrigation, planting, harvesting, feasts, festivals, ceremonies, dances, rituals, cremations, and so forth. For these sedentaries, a superbly intricate chronology rules everything in such manner that they are the ones who dominate time, thus restraining its ravages.

Where space is concerned, the Center of the World for the Balinese is the Gunung Agung, a volcanic peak 10,560 feet high, halfway up on whose slope stands the mother temple of Besakih. This mountain is Mahameru, the axis between Heaven and earth; it is likewise for them the Navel of the World, which has its correspondences with the World Tree.

Microcosmically their center is the home-and-shrine within a walled compound. The Balinese love walls, both for what they exclude, and for what they include. The village is the extension of the home, while the neighboring community is already the beginning of 'elsewhere', 'out there'. When a Balinese goes forth, he brings his 'wall' with him. I was in Bali shortly prior to World War II and had the opportunity to witness firsthand the remarkable insularity of these people. Nowhere else in the world have I met a humanity so spontaneous, mirthful, and carefree. And nowhere else have I met a humanity more impassible, dignified, and existentially abstracted from the accidental or unconsecrated aspects of the outward ambience. There is a quality in the Balinese character best understood archetypally. The wall ever-present is that of an impersonal benevolence turned inwards towards a Center barely concealed; the aura is one of aristocratic refinement molded by the Invisible, from which the world 'out there' receives its *entire* legitimation. (Fig. 3)

At the opposite extreme from the nomadic seafaring Polynesians, the sedentary insular Balinese dread the ocean, it being for them *tenget*, malefic—a hostile, alien world, the *lower* waters to be avoided; their 'maritime' activity is limited to offshore fishing, and to propitiating the shadowy spirits inhabiting seaside temples. Bali with its mountains housing the gods is the center for these people, the rest of the world connoting progressively outer space and darkness. I once showed a Balinese youth a map of the world, on the scale of which Bali appeared as little more than a pinpoint.

'Too bad!' he said.

'What is too bad?'

'Too big!' he explained, brushing his hand across the continents and seas.

Now I chanced upon Bali as an adventurer in his teens not overly concerned about philosophy and ultimate issues. And yet before a week had passed there was growing within me an unfathomable conviction that this place was somehow neither of the East nor of the West, but strangely central to both, where time is rhythmical and matter translucent. That through this translucence flowed an intercourse between gods and men, Heaven and earth: this intuition came spontaneously and more profoundly than any artifices the reason might counter by way of denial.

A young Balinese peasant once showed me the stars brilliantly reflected at night in a flooded ricefield. He explained that this was really a vision of the flowers growing in Heaven, from where Bali itself appears in the same light. Thus did I receive my first lesson in metaphysics from an untutored youth, whose remark strikingly echoes the *Emerald Tablet* of Hermes Trismegistus (of Egyptian antiquity):

> That which is below is as that which is above,
> And that which is above is as that which is below.

*

To be a Christian missionary in Bali is like a cat scratching on glass. Several years prior to the time of my visit a man or two of the cloth had been allowed on the island to practice his calling. A convert was finally made, through the pretext that becoming a Christian would absolve him from taxation. When he found this to be inexact, he consulted with the village priest, who counselled the duped convert to kill the missionary. This was done, after which episode (and previous ones of a similar pattern) the Dutch administration at Den Pasar forbade further mission work.

The final confrontation between the Netherlands and Bali came in the *puputan*—'fight to the finish'—of 1906, a massive bloodbath which was mostly suicidal on the part of the proud Balinese, who chose death over subjection. But the silver lining to this black episode manifested in the form of the 'protectorate' subsequently established by the chastened Dutch, whose experience had taught them that a policy of *laissez-faire* was in the best interests of both parties. As a consequence the Dutch wisely put their own form of wall or sort of 'magic circle' around the island, to the end that the traditions within be preserved and encouraged, while the world without be maintained, precisely, in that excluded category. Even the flux of tourism was regulated in a manner to see that the contact between natives and foreigners be kept virtually inexistent.

A Balinese priest once assured me that Jesus Christ would be welcome to a shrine here, although the Christian Deity could not supersede the other gods, who own the island, they having been the first to arrive. He also said, 'Our land belongs to the gods, who have simply put us here as caretakers'—a perspective which ties in with the Qur'ânic concept of vicegerent alluded to earlier in this chapter.

Three key words lingered on the lips of every visitor of discernment and intelligence returning from pre-war Bali: *dignity, beauty, happiness*.

4 Realism to Nominalism: The Watershed

> Belief in the unity or eternity of matter is incomprehensible; and only common, worldly-minded people, for purely materialistic reasons, covet this hypothesis. *Prajñâ-Pâramitâ*[1]

THE STRESS in the preceding chapter has been on the intimate continuity and reciprocity between subject and object common to early humanity; and several examples or 'contours' of this primordial norm were instanced at random concerning societies that have managed at least partially to maintain this inward/outward identity up to the threshold or more of our present century.

Certainly there are other civilizations, each with its own characteristic genius, that have retained well into the past one hundred years even, some of the perennial reverberations of their relative origins; countries like Japan, Tibet, India, Morocco, and certain sectors of Eastern Europe as regards the Orthodox Church come to mind by way of example.

The focus now is to be on the opposite phenomenon: on the loss of this perspective of reciprocal inward/outward polarity, and on the concomitant anatomy of separation and eventual deviation. The inception of the shift in viewpoint most directly affecting us as Westerners is to be found in the classical world of ancient Greece and Rome. It is a world with which we can identify, since its outwardly oriented humanism and secular historicity—as contrasted with the cyclical or atemporal outlook common to all traditional societies, and

1. Mahâyâna Buddhist sutra; tr. from the Chinese by William Gemmell, London, 1912.

which Mircea Eliade calls the 'Eternal Return'[1]—speaks to everything that modern man feels comfortable with in the word 'enlightenment'.

Christianity it is true is seen by this optic as having reversed the shift in viewpoint, thus 'atavistically' relegating civilization's advance by some thousand or more years into a limbo complacently dubbed, particularly since the eighteenth century, as the Dark Ages, a 'Ptolemaic' ignorance to be dispelled only by the 'Copernican' Sun of the Renaissance: at least this comes close to how our standard history books with their progressist mentality portray it. The legacy of the 'Dark Ages' from there on would be the lot of Islam, and to a lesser extent of the Eastern Church, to maintain.

More realistically, the Mediterranean world at the time of the Caesars harbored a chaotic ferment of oppositions—patrician wisdom counter to superstitious and idolatrous cults, Vandal vigor and rapaciousness counter to Roman luxuriance and decadence—a disequilibrium reminiscent in some ways of our present social setting, with the major difference that the people of those times retained a measure of human amplitude whether for virtue or for vice, an amplitude that is quasi-absent in modern man, now become in many instances subhuman.

The vital substance of the Roman Empire was accordingly, however disordered, potentially capable of spiritual regeneration, and this explains the proneness of that world to being catalyzed by tidings from the East heralding an extraordinary event that had the ring of Evidence, namely, the proclaimed descent of God in human form. In the course of the ensuing centuries the subjective or inner pole thus became again enthroned in men's hearts, while the objective or phenomenal pole was now regarded in an altogether different light—an illusory good to be renounced in its negative connotations as the World, the Flesh, and the Devil. All of this admittedly was at best a precarious restoration, as the history of the Middle Ages amply demonstrates; but it none the less brought a new order of things, reinforced moreover by the unprecedented sweep of Islam over the Near Eastern and Mediterranean worlds in the eighth and ninth centuries, so that even the enmity engendered between the two religions by their outwardly irreconcilable dogmas served to further the affirmation of faith on either side.

1. See Suggested Reading.

★

A comparison of the Byzantine Christian art of sixth-century Ravenna with that of the twelfth- and thirteenth-century Venetian style affords a glimpse of the transition from the Roman to the Mediaeval perspective. The mosaics in S. Apollinare Nuovo and S. Vitale at Ravenna still have in their somewhat rigid but lifelike realism an air of the individualistic naturalism found in the humanistic portraits on Roman tomb paintings, whereas the post-Iconoclastic work in the Basilica of San Marco at Venice and the Cathedral on the neighboring island of Torcello now has in its luminous and childlike spontaneity and 'innocence' a mystery of prototypal otherworldliness which is purely Christian.

It was the early ages of the Church that begot the great thaumaturgic and apotropaic saints—monumental figures like George,[1] (Fig. 4) Anthony, Christopher, Barbara, Ursula, Catherine of Alexandria—archetypal presences from the subjective or inward pole perennially influencing the workings of the objective or outward pole while manifesting so little of their human traces on the latter as to leave modern-day skeptics doubting of their historical reality.[2]

St. Nicholas is a case in point. The facts at best are scanty: during the time of the emperor Diocletian (reigned 284–305) there was a bishop of Myra in Lycia, southwest Asia Minor, named Nicholas; he was imprisoned, but later attended the first Oecumenical Council of Nicea in 325, although no records attest to this; the first known church witnessing his cult is that of Saints Priscus and Nicholas built at Constantinople by the emperor Justinian (reigned 527–565). The Greek Nikolaos means 'Victorious Army', and his Feast Day is December 6. So much for the record; verifiable sources for the essentials constituting his veneration are untraceable.

Yet he is the patron saint of Russia, Greece, Naples, Sicily, Apulia, Lorraine, Limerick, just to start a list. Prior to the time of America's

1. See our 'The Dragon that Swallowed St. George', *Studies in Comparative Religion*, Summer 1976.

2. Curiously enough, the Bollandists with their 'science' of hagiography have no compunction in excising from their *Acta Sanctorum* any saint – however venerable the cult surrounding him – who lacks a sufficient 'fossil record', so to speak; whereas the evolutionists with their 'science' of transformism have just as little compunction in accepting an unverifiable thesis like that of ape-into-man – and perish the lack of evidence!

Figure 1

Balinese temple with split gate (the northern temple Meduwe Karang at Kubutambahan).

Figure 3

Emerging from the Night of Beatitude onto the stage of time (Balinese dancer).

Figure 4

ST. GEORGE. Prototype of the great apotropaic and intercessionary saints of the early church.

Figure 5

KWANNON, Japanese

discovery there were 385 English churches dedicated to St. Nicholas, compared with 202 to George, England's patron saint. A contemporary census of Greek churches records 359 dedicated to St. Nicholas, only the Virgin, with 752, having more. Of the fifty-some central windows in Chartres Cathedral, four exclusively honor the celebrated Bishop of Myra.

His prelacy was of a secular cast and thus outside the spiritual model exemplified by cloistered ascetics. Moreover, he was not a martyr, except in the etymological sense of *witness*. But conversely his communion with the world made him the ideal patron of children, scholars, merchants, sailors, travellers, soldiers, knights, and prisoners—not to mention those following the pedestrian callings of coopers, bakers, pawnbrokers and brewers.

Nicholas was a myroblite, that is, a saint whose relics and shrine emit a sweet-scented myrrh or balsam of restorative and grace-giving properties. His tomb in Myra thus became a renowned center of pilgrimage for Christians and Saracens both, and so would it have remained had not the people of Bari in southern Italy—anticipating a similar intention on the part of the Venetians—forcibly translated his body to their hometown in 1087, where the tomb stands to this day. St Catherine of Alexandria, patroness of philosophers and likewise a myroblite (now removed from the *Acta Sanctorum*), whose relics are at Mount Sinai and Rouen, was closely associated with St. Nicholas in the Middle Ages, and serves in some way as his intellectual 'consort' or—as the Hindus would call it—*shakti*.

The favors of St. Nicholas were not lost on mediaeval Jewry, through his aid to one of their faith. Moreover few people today realize that the three golden balls seen hanging above the entrance to pawnshops—a custom originating with Lombard pawnbrokers—commemorate the event of an impoverished father clandestinely receiving a bag of gold from the saint as dowry for each of his three daughters to save them from a dissolute fate—an image perhaps adopted to offset mercenary associations with gold.

The axial meaning of the Christmas tree was mentioned in the preceding chapter. For those to whom symbolism is more than just a dead letter, and who understand that everything important to ancient man had a signification, the association of St. Nicholas with gifts by the fireside at the season of Christ's birth will not seem fortuitous, since the hearth originally stood for the navel of the earth and the chimney the vertical axis for the pneumatic concourse between man's offerings and Heaven's response. The function of St. Nicholas when viewed in this light and isolated from the profanation born of superstition and ignorance thus reveals itself as initially a personification of the workings of the Holy Ghost, understood in an axial or *polar* context.

To see the luminary saints of early Christianity in a meaningful perspective, they can best be comprehended as recurring human incarnations of archetypal essences radiating graces on their devotees in much the manner of the Bodhisattvas in Mahayana Buddhism, such as Maitreya, Mañjuśri, and Avalokiteśvara. These tutelary figures associated with the Bodhisattva Dharmâkara, venerated in the Pure Land School of Amidism as the Buddha Amitâbha, are accepted in Japan, China, and Tibet as 'historical' personages on the word of the Buddha Śâkyamuni, although they belong to a world system antedating ours by ten kalpas (figuratively, some 43 billion, 200 million solar years). Easterners, in other words, do not have a problem with a sequence that is purely causal and atemporal, and which can only be resolved or grasped in the transcendent immediacy of the Supreme Principle which is totally situated outside of the duality 'before-after'. Amitâbha moreover belongs to a category of Buddhas that are *anupapadaka* 'without parentage', and for which we have similar examples in the Hebrew Melchizidek and the Moslem al-Khiḍr.

St. John of Damascus (d.750), a doctor of the Church, is recorded in the *Encomium Methodii* of Migne's *Patrologia Graeca* as saying that 'it is easier to count the waves of the sea, the drops of rain, the stars, and with a glance see all the Atlantic than to recount in detail God's marvels accomplished through St. Nicholas'.[1]

And below are excerpts from a homily of St. Peter Damian (d. 1072), Cardinal Bishop of Ostia and likewise a doctor of the Church, on the Feast of St. Nicholas, as given in Migne's *Prologia Latina*:

> Here is Nicholas, whose miracles spread throughout the world, whom the whole earth and those who live on it praise. So many and so astonishing are the known miracles that all the cunning of poets does not suffice to describe them. . .. He is glorified on the sea, praised on earth, invoked in every danger. . ..

1. Charles W. Jones, *Saint Nicholas of Myra, Bari, and Manhattan*, Chicago, University of Chicago Press, 1978, p. 73.

> Don't think me oratorical, or loading my words with gilded commonplaces; for the whole world feels the benefactions of that confessor. Pilgrims to his holy corpse from the ends of the known world testify to the miracles they have seen wrought through him. Not only among Christians but even among pagans such authority speeds to the invocation of his holy name that they eagerly become one in praise and glorification of his holiness.[1]

And now for comparison are some words of the Buddha enumerating the graces to be had through holding to the name of Avalokiteśvara. These passages are taken from Chapter 25 of the *Saddharma-puṇḍarika*, 'The Lotus of the Good Law', as rendered in the Japanese *Kwannon Sutra*, where Avalokiteśvara has the name Kwanzeon Bosatsu. This sutra is particularly popular in Japan with the Zen, Tendai, Shingon, and Nichiren schools of Buddhism:

> Even when people fall into a great fire, if they hold the name of Kwanzeon Bosatsu, the fire will not scorch them because of the spiritual power of this Bosatsu. When they are tossed up and down in the surging waves, if they pronounce his name they will get into a shallower place. . ..
>
> When again a man, whether guilty or innocent, finds himself bound in chains or held with manacles, he uttering the name of Kwanzeon Bosatsu will see all these broken to pieces and be released. . ..
>
> Or if a man should encounter a party of Rakshasas, or Nagas exhaling poison, or evil spirits, let his thought dwell on the power of Kwannon, and no harm will ever be done to him. . ..
>
> [Kwannon] (Fig. 5) is a pure, spotless light and, like the sun, dispels all darkness with wisdom, and also subverts the disastrous effects of wind and fire; his all-illuminating light fills the world. . ..
>
> He is in possession of all merits, regards all things with an eye of compassion, and like the ocean holds in himself an inestimable mass of virtues. For this reason he is to be adored.[2]

1. Ibid., idem. pp. 159–60. This exhaustive work of research dedicated to the end of dismantling a 'myth' is the source for some of the above documentation on St. Nicholas.
2. Daisetz Teitaro Suzuki, *Manual of Zen Buddhism*, London, Rider, 1950.

What results from these passages is the parallelism of function—whatever the difference in degree—between a St. Nicholas and an Avalokiteśvara, with the proviso that the West in regard to its archetypal saints and Archetypes in general has lost sight of the universality of dimension involved.

*

One of the 'casualties' foisted on the classical world by what Daniel Boorstin calls 'the prison of Christian dogma' was the supplanting of its cosmographic advances—spherical earth and all—with a return to the flat-earth perspective.

Now following in the footsteps of his learned 'round-earth' predecessors, namely, the Alexandrian geographer Eratosthenes (276?–?195 BC) and the Greek astronomer Hipparchus of Nicaea (*c*165–*c*127 BC), Ptolemy (Claudius Ptolemaeus of Alexandria, second century AD) had produced in addition to his mathematical treatises three major contributions to our knowledge of the world and its cosmic prolongations: his *Almagest* on astronomy, his *Geography* on cartography, and his *Tetrabiblos* concerning astrology. This trilogy of learning was a mainstay of the cultural orientation throughout the Middle Ages. In astronomy the heliocentric system of Nicolaus Copernicus (1473–1543) would eventually replace Ptolemy's geocentric view, but in cartography by contrast, the Alexandrian's furtherance to the grid pattern of latitude and longitude measurement is still valid and used today.

Yet this did not deter others coming after Ptolemy from upholding what they considered the Christian alternative of more subjective (namely, qualitative and interior) viewpoints in contradistinction to the more objective (namely, quantitative and exterior) perspectives that had inexorably been gaining the ascendancy. Thus we have the instance of Firmianus Lactantius (*c*260–*c*340) of Africa, called the 'Christian Cicero', and tutor to the son of Constantine the Great, denouncing the spherical idea of the earth out of deference to scriptural authority—for what could biblically justify the concept of people upside down with their feet pointing at ours (antipodes)? Then came the sixth-century merchant and traveller, Cosmas of Alexandria, whose early voyages reputedly took him as far as western India and Ceylon (whence his sobriquet Indicopleustes—Indian Sailor) before his conversion to Christianity and subsequent vocation as a monk in a

Sinai cloister. It was here that he wrote his *Topographia Christiana*, blasting the idea of the earth's rotundity (something he none the less either already knew, or could hardly have failed to suspect on his extended voyages) and vindicating the structure of the world according to scriptural interpretation.

What have been called 'Ecumenical' maps (from their pretension to entirety) or 'T–O' maps (from their representation of a circle enclosing a 'T' shape) survived into the fifteenth century. A typical example is found in Isidore of Seville's (560?–636) *Etymologies* (Fig. 6). East

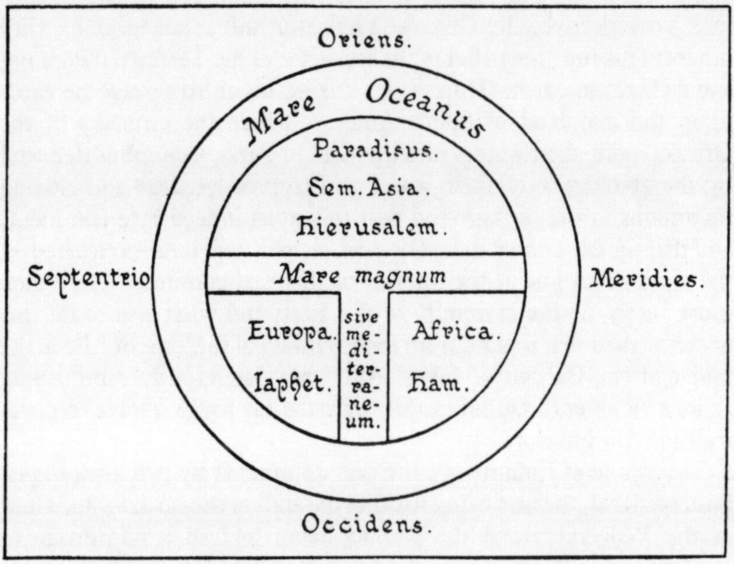

Figure 6

is placed at the *top* of the circumferential band named in Latin *Mare Oceanus* 'Ocean Sea', which girdles the land; the straight band forming the crossbar of the 'T' within is labelled 'Great Sea', with the vertical portion reading 'or Mediterranean'. Contained in the angle of the 'T' to the North is Europe, considered to be the land of Noah's son Japeth, while to the South is Africa, belonging to Ham. Just over the 'T' and placed at the center of the world is Jerusalem. Above this comes Asia, apportioned to Shem. Lastly, at the summit, is Paradise.

The whole thing is purely symbolical and symmetrical without any pretension to topography. There are also maps showing the four rivers flowing from Eden, such as the Indus, the Nile, the Tigris, and the Euphrates.

Christopher Columbus (1446?–1506) had the global concept of the earth, and with it the practical vision of a westward passage to the Orient based on information conveyed to him by the Florentine cosmographer Paolo del Pozzo Toscanelli (1397–1482), himself in turn convinced by what Marco Polo in the fourteenth century recounted of his travels, that Cathay and India could be reached from the West. Yet despite this new and strictly geographical (though Ptolemaic) perspective, the Genoese navigator still retained along with others of his time the belief in the existence of the Terrestrial Paradise somewhere on earth. Thus when during his third voyage he came upon the mainland of South America where the estuaries of the Orinoco pour their waters into the Gulf of Paria, Columbus deduced by the abundance of fresh water and tropical verdure and soaring mountains in the background that this must indeed be a continent, and that the earth must in fact be pear-shaped with its apex situated on the equinoctial line at the interior of this vast continent. Here then must surely be the extremity of the East; and what else could this water be than that issuing from the Fountain of the Tree of Life in the midst of the Garden of Eden! And only dangerously diminishing rations along with failing health thwarted his lively resolve to penetrate into the interior.

The whole of Columbus's life was dominated by two aspirations, both spiritual, the first being to unite the ends of the earth for the glory of the Redeemer, and the second being to lead a pilgrimage to Jerusalem in order to rescue the Church of the Holy Sepulchre from the hands of the Infidels. And it was in the name of the Holy Trinity that he embarked on the uncharted Atlantic with nothing but astrolabe and compass to supplement the visionary charts of Toscanelli.[1]

★

1. See Washington Irving, *Columbus – His Life and Voyages*, London, 1828. The epistle Columbus wrote to Ferdinand and Isabella on the presumed proximity of the Terrestrial Paradise is found in *Select Letters of Christopher Columbus*, tr. by R. H. Major, London, 1860, pp. 140–47.

It is a common assumption that the 'flat-earth' people of the 'Dark Ages' were 'dumber' than the 'round-earth' people of the classical world, and that the culprit in this perceived weakening of man's intelligence was religion, through its espousal of 'faith' over 'reason'.

But there are other factors not to be overlooked in correctly assessing the pros and cons of this assumption. For what it really comes down to is a matter of options and priorities, of whether the higher value is to be attributed to the subjective, qualitative, and inward pole, or to the objective, quantitative, and outward one. Not that the two views are incompatible, as the cosmic system of Ptolemy admirably demonstrates in its harmonious equilibrium between physical fact and spiritual truth. But not all people are gifted with this capability for integral vision, as Schuon has observed in his remark already cited in the previous chapter about 'the human incapacity to exert the intelligence on all planes at once'.

It would in any case be presumptuous on our part to set limits to the potentialities of the human intellect of one thousand years ago, and all the more so as the Intellect in its pure form is a timeless universal faculty which descends into without being determined by—even if circumstantially more or less veiled by—the limitations of the individual human state, whose domain is that of the reason. The rational faculty is dualistic; the Intellect is the otherworldly eye of single vision, which explains its representation in Christian and Hindu iconography as a frontal eye.

Be this as it may, the Church authorities in the early Christian era could well have opted for a 'spiritual cartography' of a symbolism propitious to man's final ends, rather than a purely physical and factual cartography which in most instances would be neither needful nor expedient in the *sedentary* world that was Christianity. Too much preoccupation with the details of distant geography could even be a Pandora's Box to Christian sedentaries who had no business abroad. History indeed has shamelessly shown us how the lofty ideals of Columbus were to be replaced in a matter of short years by a deviated sedentary rapaciousness scattered over the seven seas, and which has never since abated. Not without cause was it considered forbidden in earlier times to breach the Pillars of Hercules.

To say that the ancient Church authorities were dupes of a 'flat-earth' perspective is rather on a par with attributing ignorance of the wheel to those civilizations that have for their purposes forgone its use—this attribution betraying a cavalier attitude which tells us more about the pride of the wheel-driven mechanic than it does about the mentality of the 'wheel-less savage'.

In Hindu cosmography, Mount Meru with its summit in Paradise (Ilâvṛita) dominates the center, from which the emanating worlds descend in seven concentric rings, our final earth resting on the tortoise Akûpâra. Before sharing the scorn that Lord Macaulay manifested in the face of such 'fancies', one should realize that the point of departure here is mythological, with spiritual truths in view. The tortoise, for example, in its ability to interiorize and exteriorize at will, symbolizes primordial man at the threshold of inner and outer, manifestation/nonmanifestation, who in the role of vicegerent, as we have seen, is invested with the responsibility for upholding cosmic order.

It is important to know moreover, that parallel with this view, the Indians accepted the findings of Greek astronomy concerning a spherical earth and measurements in longitude and latitude.[1] They were, in other words, capable of assimilating simultaneously a plurality of world pictures addressing the multiple channels open to the human intelligence and intellect.

There is, in fact, a disagreeably condescending note in the expressions 'flat world' or 'geocentric system'—terms that are exploited as 'buzzwords' calculated to set automatically into motion a feeling of superiority with us over our ancient forebears. After all, our daily existential experience is of a flat (static) earth, and our architecture and road-building, among innumerable examples, would be disastrous without it. A round (dynamic) earth for most practical experience is mainly a theory that we carry in our heads when we happen to think upon it, and it is only in the global and extra-terrestrial extensions of contemporary technology that the concept is really operative. Even the maps and charts we use are mostly Mercator, or flat, projections dating from the middle of the seventeenth century, while present-day Global Air-Age Maps are terrestrial planisphere or gnomonic projection charts with the North Pole at the center, a concept introduced in 1696; all these charts have the convenience of rendering spherical extension on a flat surface. The matter of geo- and heliocentric positions will be expanded in the next chapter.

1. See the article on Cosmology in Benjamin Walker's *Hindu World*, London, Allen & Unwin, 1968.

Paradoxically, the very persons most prone to mock the flat-earth perspective are the same ones most prone to feel perfectly complacent with what we have elsewhere termed a 'one-world syndrome', namely, the belief that our physical universe is the sole reality that exists, without any reference to higher orders of being. We could just as well charge these people with having a 'flat-world syndrome', since cosmologically speaking these are the people most deluded by a perspective that is purely horizontal with no vertical dimension. And they are locked into this optic beyond all reach.

Unlike Islam with its *sharī'ah* and Hinduism with its *dharma śâstra*, Christianity never had a clearly defined code integrally regulating those aspects of human conduct outside the direct province of the moral life, the 'rule of thumb' being pretty much 'render unto Caesar the things that are Caesar's'. This is why sedentarily oriented Christians could breach the static limits of their world and repeatedly set out with impunity on distant voyages every time something new was in the air to be discovered; and all the more so as the pretext (forget the lure of gold and the rest) was invariably enhanced with the lofty asseveration of gaining more heathens for the Gospel faith.

Shown in the preceding chapter is the aversion Balinese Hindus have for personal displacement. This attitude has its roots in the sedentary nature of Hinduism, where Indian Hindus are discouraged from travelling even outside of their own province or state, and where overseas voyaging in particular is interdicted as being spiritually corrupting; and the *Baudhâyana Dharmaśâstra* details the severe penances to be undergone by brâhmins transgressing this prohibition. Despite—or perhaps before—all this, the ancient Indians notwithstanding reached most quarters of the eastern hemisphere, Indonesia being a case in point.[1]

The issue at stake is one of allegiances. As the pole *subject* following the definition given it throughout this book is hierarchically in essence the origin of the pole *object*, and never vice-versa, it is normal for man to hold to the former and not allow his natural instincts to gravitate him towards the latter, which is eventually illusory, so that Narcissus-like he ultimately becomes the pawn of mere semblances. Yet as our sublunar world exists by the play of polarities, both subject and object are indispensable to our well-being as mortal creatures; and the art is

1. See the entry on Travel in Walker, op. cit.

to find a working equilibrium between the two poles. It is from the objective pole that we receive our bodily sustenance,[1] our soul-refreshment (from the splendors of creation), and our spiritual supports (theophanies) for contemplation. Yet in any civilization where the subjective pole predominates, the objective pole is going to be relativized, as has been demonstrated in the matter of mediaeval cartography. After all, man is essentially a spiritual creature, and not just a machine (as in the 'flattering' idiom of G. I. Gurdjieff), whatever people today may believe.

Even the vaunted marvels of technology are in fact coming purely from the subject's manipulation of the object; the object (without reference for the moment to any animic elements) counts for nothing in this except as malleable substance. And only the Subject in the meaning intended here can be Transcendent, never the object. At most the object can symbolize Transcendency, such as is mirrored in the illimitable vastness of the outside universe.

Confronted with the polarity inner-subject/outer-object, man has the choice of several responses or alternatives which are legitimate with regard to his spiritual ends. He can elect the life of the cloister over the life of the world as an attitude which corresponds to the needs of his inner nature. Or he can elect the householder's life over the monastic ideal as a means of praising God more outwardly through His manifold traces in creation.

With Christianity, as with Buddhism, the accent is on other-worldliness, the outside world being essentially regarded as illusory, deceptive, and ultimately negative. With Hinduism and Islam by contrast, the traces of the Divine wherever found in the universe are seen as potential spiritual supports. But this latter point of view also manifests in the Chivalric and Hermetic dimensions of Christianity, as in the tantric dimensions of Buddhism. And the compensatory balance to the theophanic aspect of Hinduism and Islam shows itself through a current within these two religions of uncompromising

1. Although not in an absolute way, as is demonstrated in the case of the celebrated Swiss lay saint, Nicholas of Flue (1417–87). Soldier, patriot, prosperous farmer, and father of ten, whose mediation at the Compromise of Stans preserved the menaced Swiss Confederation, he was above all a venerable hermit. And what most awed his many visitors that came for counsel was the extraordinary gift he had for fasting from food to a degree that was miraculous. There are other cases, Catherine of Siena (also actively engaged in politics) being one, of saints subsisting largely on the Eucharist.

ascesis. The whole matter really comes down to spiritual realization through total inclusion of the multiform content of being ('All is *Âtmâ*') as contrasted with the equally legitimate position of realization through total exclusion of all 'otherness' (the Hindu *neti, neti*: 'neither this nor that', and the Islamic *Shahâdah: Lâ ilâha illa'Llâh*, 'There is no divinity but the Divinity'—a formula that can also be used inclusively).

★

The continuing friction between Pope and Emperor during the Middle Ages was in no small part due to the conflict of interests between the two poles in question. As Pontifex—a title taken from Roman times—the Holy Roman Emperor of whom Charlemagne (742–814) is the exemplar, (Fig. 7) saw himself as Mediator between Heaven and Earth for the well-ordering of society. As Pontifex Maximus, the Pope saw himself as Mediator between Heaven and Earth, not only for the salvation of souls, but also for the prerogatives of worldly dominion. What Dante saw in all this was a contest for supremacy, and he detailed in his *De Monarchia* the need for the spiritual authority on the one hand and the temporal power on the other to function less hierarchically, the first over the second, and more collaterally each in its own domain.

But the question was one of defining domains. In the absence of a clearly formulated social code, the Church by the legitimate nature of its otherworldly message was nevertheless imprudently led to underestimate the rightful needs of Caesar, tending to view his constituency as what we have already seen was considered that of the World, the Flesh, and the Devil. This too summary dismissal of the objective pole of existence could on occasion leave the Emperor with the responsibility to step in as the protector and patron of the universities, the guilds, the arts, sciences, and in short all that constitutes the licit outward riches there for the nourishment of the human soul.

Vikings christianized with an eye to the best of both worlds, on crusades to the Holy Land, kept in form by plundering the European coastal cities that lay along their passage. But when they came to Palermo they dropped anchor, left weapons aboard ship, and embraced their host, Roger the Norman (= Norseman = Viking), at whose court they feasted and regaled prior to the final lap of crusading. For Roger II, King of Sicily (reigned 1130–54), exemplifying the 'rightful needs of Caesar', had one of the most sumptuous kingdoms in Christendom, where church and mosque stood side by side, and Greek scholars debated by the hour with Moslem philosophers and Christian savants. No doubt Jewish Kabbalists were also in attendance, appearing as they did in southern Europe around this time. Here at Palermo by the King's invitation the celebrated Arab geographer Ibn Idrisi composed his famous *Al Rojari* which was then the foremost work in Arabian geography, and he presented Roger with a celestial sphere and likewise a disk—both in silver—depicting the accumulated geographical knowledge of the day. This court was also home to the historian Nilus Doxopatrius.

The King in deference to the usage of his Viking and Arabian company kept a harem; and the liveliness, the cosmopolitan commerce, and the protean activities of this enlightened court belied the fact that it was buried in the 'Dark Ages'. Indeed, had Roger and his courtiers been apprised of this circumstance, they would surely have honored it with a toast to its perpetuity. Now whose effigy graced the coinage of the realm? Whose else indeed but St. Nicholas's!

Roger's grandson, Frederick II (1194–1250), saw to it between military campaigns that the court at Palermo lost none of its lustre: he likewise kept a harem, with eunuchs, at Lucera, and he too surrounded himself with Jewish, Moslem, and Christian scholars. He added for the rest a large menagerie of exotic animals, such as white elephants acquired on a crusade to Jerusalem.[1] He also founded the University of Naples, and became renowned as a lawgiver. Italian poetry according to Dante had its birth in his court.[2]

This Emperor was known as *stupor mundi* 'Wonder of the World', and after his death a popular legend formed about his reposing

1. Frederick embarked on this crusade despite the fact that he had been previously excommunicated for insubordination by Gregory IX in 1239. Hence we have the paradox of this King paying his respects to the Church of the Holy Sepulchre while followed by a priest pronouncing anathema on the precincts considered profaned by Frederick's presence – anathema thus being pronounced on the stones of the holiest site in Christendom! The Emperor also mounted the minaret of a mosque to hear the Islamic call to prayer, but forbade the priest to follow, saying, 'This is not for you!'

2. That the papacy also had its share of illustrious statesmen and scholars is attested to in the person of Sylvester II (Gerbert), pope from 999–1003, the most accomplished scholar of his age in mathematics, literature, music, and astronomy – this latter being studied on terrestrial and celestial globes.

dormant in a cavern in the Kyffhäusser mountain, against the day when he would arise and restore to the Empire its Golden Age.

★

Realism in the 'Dark Ages' meant the exact opposite of what the word connotes today. The Schoolmen or Scholastics, who had their start in the establishments of learning founded by Charles the Great, the first Holy Roman Emperor, used the word to mean that all seeming reality in our world is entirely infused by the sole ultimate Reality of the Universals or Ideas as propounded by Plato. These are the Informing Essences, Archetypes, Exemplars, and Qualities of which it has been a question in Chapter 2. The Scholastics taught that the Universals are *ante res*, *in rebus*, and *post res*, meaning that their existence is prior to, within, and subsequent to outward things. The Schoolmen insisted in effect that Realism is 'an assertion of the rights of the subject', and this axiom shows the primacy accorded to the subjective pole over the objective during the mediaeval period.

Realism in later times has come to signify the notion that matter and sense objects have a concrete reality in their own 'right', they being considered to embody a true existence independently of Ideas—a position which could certainly be called 'an assertion of the rights of the object', following the dictionary definition of the word *object*'s popular use as given in Chapter 1.

In order to distinguish between usages of the term *realism*, the first (original) meaning sometimes goes by the name of *logical realism*, and the second (modern) meaning by the name of *epistemological* or *natural realism*.

Nominalism is the term that describes the view taken by the opponents of Scholastic Realism, namely, those who say in substance that there are no universal essences, a stand which comes down to the realism of our day.

The term *conceptualism* has been applied to an intermediary stance between the other two postulates, and need not concern us here.

Nominalism is mainly associated with the teaching of the English Franciscan theologian, William of Ockham (*c*1285–1349), but already before him the tendency to take the concrete fact for the final reality manifested in a person like Roscellinus of Compiègne (*c*1050–*c*1122), generally considered the first Nominalist by his teaching that the three

Persons of the Trinity are necessarily separate entities (Tritheism). And even before Roscellinus there was Berengarius of Tours (d.1088), who attacked the doctrine of Transubstantiation on the basis that reason had to adjudicate in matters of faith, and according to reason the Eucharist could only be understood as a figure of Christ. For how could the body of Christ be literally present on many altars at the same time? The amazing thing is that these men got roughed up, but not condemned, by the Church for their assaults on orthodox theology.

And even antedating Berengarius was the ninth-century Aquitanian monk, Ratramnus, who probably holds the record as the first Nominalist for his assertion that the Eucharist is merely a 'memorial'—a postulate that would have to bide its time more than a thousand years before being officially promulgated by Vatican II.

One might reasonably approximate the first real stirrings of Nominalism in the West to the time of the Great Schism between the Eastern and the Western Church, the key date being 16 July 1054, and the pretext an altercation over the *filioque* clause in the Nicene Creed.[1] It was on the afternoon of this day while Patriarch Michael Cerularius was presiding over a Mass in the Hagia Sophia of Constantinople that Cardinal Humbert, Bishop of Silva Candida, accompanied by two other cardinals, laid on the altar of the basilica a document of excommunication from Pope Leo IX, thus in effect anathematizing the patriarch, and with him the whole ecclesiastical and temporal hierarchy from Emperor down to the last religious under the Patriarch's authority.

What was really at the root of the matter was the mounting tension between a more rational and dynamic world—the Roman sphere, and a more contemplative and static world—the Eastern sphere. Constantine the Great (280?–337), the Emperor who made Christianity the religion of the Empire, had in fact chosen for his capital the ancient Greek city of Byzantium, which he then named Constanti-

1. The issue is not merely academic. The original Nicene-Constantinopolitan Creed, still observed by the Eastern Church, reads as follows *except for* the portion in italics which is the *filioque* clause that has been added by the Roman Church: 'I believe . . . in the Holy Spirit, the Lord, the Giver of Life, who proceeds from the Father *and the Son*, who with the Father and the Son together is worshipped and together glorified.' This addition in effect relativizes the Absoluteness of the Supreme Subject, which from the standpoint of this book is already a big step towards exteriorizing the objective pole of existence.

nople, dedicating it to the Virgin, a reason for his choice being to establish the Church in a 'New Rome' that had no antecedents in a paganism of the type rampant in the old Roman world.

*

A document which caused the very mischief between East and West, Papacy and Empire, was the infamous Donation of Constantine so called (*Constitutum Constantini*), composed around 800 AD. The British historian, Viscount Bryce, considered it 'the most stupendous of all mediaeval forgeries'. Its fraudulent character was only brought to light in the fifteenth century by Nicholas of Cusa and Laurentius Valla.

In this 'Edict' Constantine, out of gratitude to Sylvester I (Pope 314–35) for curing him of 'leprosy', would have conferred on this Pope and *all his successors* dominion over East and West, thus making Rome the political—not to mention spiritual—linchpin of the entire Christian world. So scandalized was Dante by this supposed betrayal of the principle keeping separate the functions of spiritual authority and temporal power, that he bitterly chastised Constantine in the *Inferno*:

> Ah Constantine! to how much ill gave birth,
> not thy conversion, but that dower which
> the first rich Father took from thee! (XIX.115–17).

Could this Emperor have answered from his grave, he certainly would have remonstrated against the poet's too ready credulity to think that a ruler of Constantine's stature could be so naïve as to betray Heaven's trust of rulership. As for his 'leprosy', he would have explained it as an *image* of the soul that he presented to Sylvester for purging and healing.

But Constantine did not speak from the otherworld, and Dante remained haunted by the terrible incongruity of this treacherously clever document. The problem which Dante clearly saw is that the 'outer' was gaining ascendancy over the 'inner', and he kept coming back to this issue in his writings. Thus in the *Purgatorio*:

> . . . the sword is joined to the crook; and
> the one together with the other must
> perforce go ill;

> . . . the Church of Rome, by confounding
> two powers in herself, falls into the
> mire, and fouls herself and her burden.
>
> (XVI. 109–11, 127–29).

And in the *De Monarchia*: 'Constantine had no power to alienate the imperial dignity, nor had the church power to receive it.' (Ch. X).

Justinian the Great (483–565) personally supervised the construction of the Hagia Sophia, the form of which was said to have been revealed to him in a dream. And the great nave, flooded with light from the soaring dome above, was conceived as the place of convergence for the 'two halves of God'—the Patriarch (subject) with his clergy, and the Emperor (object) with his court—to meet in worship.[1]

It was a mere twenty years after the edict of excommunication from Rome had been placed on the altar of this basilica that Pope Gregory VII (Hildebrand, reigned 1073–85), fortified in the assurance of Constantine's 'Gift', conceived the gigantic military campaign that was to be the basis of the forthcoming Crusades. The Pope was to lead in person a Holy War against Normans, Greeks, and Muslims, with the aim of destroying the Caliphate, taking Constantinople and ending the Greek Schism, seizing Palestine, and regaining Jerusalem and North Africa. Gregory died before he could undertake this project, and it was reserved for Urban II to preach the First Crusade at Clermont in Auvergne in 1095.[2]

The purpose of the foregoing survey—Crusades apart—is to show how a Western European mentality easily given to passional excesses was inexorably turning more and more away from the qualitative concept of a spiritual kingdom, through a mounting attraction towards the outer pole of quantitative earthly dominion and possession. If one protests that the deeds of the worst of the popes in the coming centuries pale in comparison with the atrocities committed by the worst of the sultans in the Ottoman Empire,[3] the difference is that the latter concerns more particularly excessive personal depravity, whereas the former has a more grandiose element riveted in its ambition to restructure the whole social order.

1. See Lord Kinross, *Hagia Sophia*, New York, Newsweek, 1972.
2. See John Laux, *Church History*, New York, Benziger Brothers, 1937; and Malachi Martin, *The Decline and Fall of the Roman Church*, New York, Putnam's, 1981.
3. See Noel Barber, *The Lords of the Golden Horn*, London, Macmillan 1973.

★

Although Ockham was hardly the first Nominalist, he was indisputably a watershed in the separation between 'Light' and 'Darkness' in the Middle Ages—his affiliation to which pole depending on one's reading of the matter. He followed in the wake of the last great Scholastic, St. Thomas Aquinas (1225?–74), whose *Summa Theologica* and other works endure as the consummate formulation of Realism, even if metaphysically the profoundest exposition remains that of Erigena. Aquinas none the less did leave one element of his *opus magnum* particularly penetrable to Nominalism, in that his work is centered within the analytical, discursive orbit of Aristotelianism rather than within the synthetic and unitive orbit of Platonism—a shift of outlook in Scholasticism for which the French philosopher and theologian, Peter Abélard (1079–1142), was largely responsible.

The Aleppine Sufi philosopher Suhrawardi (d.1191) in his *Hikmatu'l-Ishrâq*, 'Philosophy of Illumination', makes the distinction clear:

> I first grasped a definite truth by mystic intuition, and then sought to demonstrate it . . . whereas the Peripatetics follow the reverse method . . .. *Our* master is Plato . . . whereas Aristotle remains the loadstar of those who seek truth by the empirical method . . .. This mystic-Platonic method is a different kind of philosophy, and a shorter way than that of the Peripatetics, which loses itself in secondary questions.[1]

The Cambridge Platonist, John Smith (1618–52), deals more harshly with the Stagirite: 'He hath so defaced the sacred monuments, . . . that elder philosophy which he so corrupts'.[2] As a logician, Aristotle can hardly be surpassed, but logic still pertains to the domain of reason, and the rational faculty unaided cannot transcend the dualism of which it is composed. Certainly Aristotelianism still is intellectually oriented; it is the manner that is indirect.

For Ockham there was nothing even to transcend; God could well be 'up there', but this world 'down here' had it all. For him concepts *extra mentem* were just that—concepts, and the fewer of them the better, a way of looking at things that was to become designated by

1. In Scott, *Hermetica* (IV; p. 264 n.2).
2. *Select Discourses*, 1821 London, p. 114.

others as Ockham's Razor: *Entia non sunt multiplicanda praeter necessitatem*: 'Entities are not to be multiplied beyond necessity'. This angle of vision, known as the 'law of parsimony', may have appeared as but a small hole in the dike separating the 'upper' from the 'lower' waters, yet it functioned as a fissure opening onto the quantitative and exterior pole of manifestation, a breach that over the centuries would unloose the whole form of scientific mentality on which the modern world is fabricated.

The razor shaves off what for Ockham is the surplus fat: pare away the flesh, and the skeleton that remains=reality. This corresponds to what Alfred North Whitehead calls the bifurcation concept. Titus Burckhardt puts the matter thus:

> All true progress of natural science resides in the fact that it disengages itself more and more from subjectivity and that it brings out more and more clearly what exists independently of human conceiving, without troubling itself about the fact that the result has only a distant likeness to that which the original perception took for real. . .. It is the entire 'human conception of things' . . . that is in question; only mathematical thought is considered to be objective or true. The latter allows in fact a maximum of generalization while remaining linked to number, so that it can be verified on the quantitative plane. It does not, however, include the whole of reality as communicated to us through our senses. . .. Thus it is that those sensible qualities . . . such as colors, odors, savors, and sensations . . . are considered to be subjective impressions implying no objective quality, or possessing no other reality than that which belongs to their indirect physical causes, as for example, in the case of colors, to the various frequencies of the luminous waves. . .. Modern science therefore asks us to sacrifice a good part of that which makes for us the reality of the world.[1]

1. Titus Burckhardt, 'Cosmology and Modern Science', in *The Sword of Gnosis*, ed. Jacob Needleman, Baltimore, Penguin Books, 1974, pp. 128–29.

Burckhardt's point was brought home some years ago when a professor of the 'exact Sciences' at Massachusetts Institute of Technology was dining with a professor of the humanities at the same institution. The physicist all at once turned to the philosopher and said, 'I have been pondering on what makes the difference between you people and us, and now I have put my finger on it. It is that we *count*, and you don't!' Far more pertinent is the fact that a scientist like Erwin Schroedinger, awarded the 1933 Nobel

The fact that Ockham, the leading exponent of Nominalism, was honored by his contemporaries with the titles both of *Doctor Invincibilis* and *Venerabilis Inceptor* obliges us to surmise, either that among his peers none were intellectually his match, or else that Nominalism was already in the air simply awaiting an imprimatur. The second title for the rest hints at this, being an oxymoron in the style of 'conservative liberal' or 'traditional innovator'.

Just before Ockham's time even, the English Franciscan philosopher and scientist, Roger Bacon (1214?–92), was introducing the concept of experimental science, although strictly in the name of religion. But in striving to reconcile the world of universals with that of particulars, he fell between two stools, thus leaving his encyclopaedic writings pointing contradictorily two ways at once. And by incongruously demeaning the teachings of his foremost intellectual contemporaries, and more especially by deprecating the *Summa Theologica* as a work full of 'puerile vanity and voluminous superfluity', this together with suspicions of his dabbling in 'magic' gained him imprisonment from 1277 to 1292, the year of his death.

Guénon has pointed out that traditional sciences, being founded on metaphysical and cosmological principles, are in no way contradictory to man's ultimate ends. Whether in the mediaeval learning based on the trivium and quadrivium, or in the guild and craft initiations, or in the Pythagorean system of numbers, the sciences that were pursued followed the laws of analogy linking together different orders or levels of cosmic reality; that is to say, there was always an adequation between the subject and object, inner and outer.

It is with the arrival of a science observing empirical methods alone that Pandora's Box gets opened, for then the indispensable reciprocity between the qualitative and quantitative poles of existence is ruptured, and the movement becomes purely a centrifugal flow towards the world 'out there'.

Thus it is that along with his contributions to the more 'neutral'

Prize in Physics for his 'wave equation' which is at the heart of modern quantum mechanics, unqualifiedly takes the position siding with that of Burckhardt's just cited. See *Quantum Questions*, ed. Ken Wilber, Boston, New Science Library, 1985, pp. 81–3 and *The Philosophical Scientists*, David Foster, New York, Dorset Press, 1985. Writes Schroedinger: 'The reason why our sentient, percipient and thinking ego is met nowhere within our scientific world picture . . . is because it is itself that world picture' (Foster, p. 140).

sciences like optics and meteorology, Roger Bacon developed the exploitation of gunpowder and worked on designs for automobiles, flying machines, and mechanically propelled boats. If imprisonment was to be his lot for some of the more daring elements in his innovative spirit, he nevertheless came out posthumously triumphant, since his successors were to honor him with the title of *Doctor Mirabilis*.

It is the other Bacon, Francis (1561–1626), who grasped the torch from Ockham. Now the factor essentially determining the messages of the speculative systems proliferating from the Renaissance onwards is, from the standpoint of this book, the particular tension of the cosmic moment at each period when a new philosophy comes to light, that is, the particular rapport at any instant in the preponderance of stress between the qualitative and the quantitative pole of existence. Hence we regard the philosophers as being by their several endowments the destined vehicles, spokesmen, or apertures for ideas caught as it were 'in the air', but do not at all regard them as having produced thoughts 'out of thin air', or for that matter, out of malice, or conscious self-interest beyond a point.

Seen this way, if reference be made to the notion of 'conspiracy', the collusion in question has really to be that of the 'cosmological constellation' of any immediate time under consideration, the philosophers' roles being limited somewhat to barometers of the prevailing 'cosmic weather'. And because in the nature of things there are going to be other thinkers aware of the errors and dangers in these new systems, there are inevitably going to be other philosophical constructions intended to refute the new systems. But the problem with both sides at issue here is that a cosmically induced intellectual obscuration, allied with the prejudices provoked by the passional predisposition of Europeans to hypercerebration, practically guarantees the odds falling in favor of the side most passive to the quantitatively gravitational 'lure' of the cosmos outwards. Certainly there is never lacking the reciprocity of a qualitative current always flowing inwards towards the Center, but this aspect of the cyclical movement becomes ever more hidden as the age of the world advances; and by the evidence of their writings it has not been the common run of philosophers 'officiating' since the Middle Ages who have had the intellectual 'antennae' to sense this current.

Whatever the case, the spadework done by Roger Bacon and the 'razor work' accomplished by Ockham had already prepared a terrain propitious for the furtherance of the projects of Francis Bacon, his chief object being to harness particulars to inductive reasoning in the end of achieving universals, i.e., the idea that the sum of the parts is what constitutes the whole. His credo amounted to the notion that truth is utility and utility truth—no 'high and vaporous imaginations'; and in his conviction that the more can be gotten from the less, Bacon might be considered the first evolving evolutionist.

Thomas Hobbes (1588–1679) who came just after him, went Bacon a step better, in applying mechanical explanations to everything, mental and material. For he regarded ultimate realities as being purely matter and motion, these causing the sensations from which knowledge is derived. Thus for him mind was the stuff of matter, and thought the effluvium of motion. Here we have Darwin anticipated.

The classic would-be sundering betwixt 'mind' and 'matter', the inward and the outward, was the 'contribution' of René Descartes (1596–1650), who like Galileo (1564–1642), attempted to explain all natural phenomena mechanically. He was an exponent of the physiological theory of the English physician William Harvey (1578–1657) and the British physicist Robert Boyle (1627–91) that interpreted all vital functions on mainly mechanistic grounds. Minds and material bodies were regarded by this French philosopher as absolutely distinct substances incapable of interaction but possessing nevertheless a certain link thanks to the intermediary of the pineal gland. Bodies in this perspective result from whorls of motion becoming extension in an automatic universe operated by a deistic God who, as Blaise Pascal (1623–62) quipped sarcastically, had only to 'donner une chiquenaude pour mettre le monde en mouvement' ('give a flick of the finger to get the world going').[1]

★

1. *Pensées*, II.77.

Jacques Maritain names the aberration of Descartes Angelism, being the concept as if man were pure spirit without body, also, the interpretation of the world and ideas that results from this perspective. And yet notwithstanding, the author of the credo *cogito ergo sum* was perfectly comfortable in his conviction, shared by the Church, of being a fervent Catholic. If a lesson can be drawn from this, it is that religions while functioning microcosmically as vehicles of salvation, are none the less macrocosmically more or less inevitably swept along with the prevailing cosmic currents and unable fully to maintain their independence from the prejudices and errors of any particular period. It is an immutable metaphysical understanding alone that is proof against this vulnerability.

It fell on the shoulders of the humanist and 'prince of rationalists', Baruch Spinoza (1632–77), to try his hand at lowering Descartes' fences and bring God and Nature together again as the Perfect Self-Existent—a euphemism for a frankly stereotyped pantheism, whatever the subtleties in the mode of presentation, since it predicates a God that is entirely immanent in pure substance. *Deus sive Natura*: there is for our sage of Amsterdam no supranatural order; nature is everything, and this All is God.

Of nobler bent was the system offered by Gottfried Wilhelm Leibniz (1646–1716), to answer what he considered the defects in Cartesianism. To accomplish this he denied the reality of corporeal matter altogether, saying that material bodies are but subjective appearances. In their place he posited simple imperishable substances called monads, acting independently although interrelated in a 'clockwork' hierarchy from God, the supreme monad, on down like a play of mirrors to the humblest substance. Thus the physical world for him was a kind of teleological mechanism based on a superlative divine harmony where everything was for the best—a conception that brought him Voltaire's (1694–1778) scorn in *Candide*: 'Si c'est ici le meilleur des mondes possibles, que sont donc les autres?' ('If this is the best possible of worlds, what then must the others be like?').

In an attempt to tumble the intellectual edifices of his contemporaries the Cambridge Platonists, the English philosopher John Locke (1632–1714) countered with the empirical argument that the mind prior to receiving experience is but a blank sheet, a *tabula rasa*. So far from there being for him innate Ideas to be cognized by intellectual Recollection, there are only sensations reaching the mind from an external world that while really existing cannot communicate its substance to the mind directly. What this means is that the sensations entering each mind are strictly private experiences, although once the mind 'thinks' upon these impressions, they then become transposed a step higher, to 'ideas of reflection'. Yet since these ideas enter from the external world, they can only be known on 'presumptive trust' as being 'reasonable probabilities'.

Thus in the eyes of his followers, Locke's 'glory' was to have demonstrated the limits of our knowledge. By this view, the vaunted 'reason' of the Rationalists is found wanting of anything even resembling existence. And at what a price!

★

Given the ever increasing philosophical encroachment of 'matter' over 'mind', George Berkeley (1685–1753), Bishop of Cloyne, felt it incumbent to reinstate both God and mind in their lawful dominions; and for this he came up with the speciously incontestable and beguiling idea that *esse est percipi*: to be is to be perceived. Thus in his *Principles of Human Knowledge* did the good bishop endeavor, in a prose at once lucid and slightly prissy, to demonstrate that the whole sensorial world depends entirely for its reality on our perception of it. To be sure, God masterminds the process; still, this does not *de facto* change the world's being but a projection of *our* minds—to believe Berkeley.

Here then is the exact reverse of the evolutionist position taken up in Chapter I, which position regards the objective (material) pole of existence, not as a pole, but as an autonomous entity independent of any subjective (cognitive) counterpart. For Berkeley treats the subjective pole—reduced to mind, moreover, namely reason and sensation—as an autonomous absolute independent of any objective (material) counterpart. And this 'absolute' for him is purely rational and flatly horizontal, without trace of a vertical or transcendent element.

Hence he considers 'the Schoolmen, those great masters of abstraction,' as caught in 'manifold inextricable labyrinths of error and dispute which their doctrine of abstract natures and notions seems to have led them into. What bickerings and controversies, and what a learned dust have been raised . . ..' (*Principles*, Introd. no. 17). Rejecting matter in any form, Berkeley casts derision on 'that antiquated and so much ridiculed notion of *materia prima*, to be met with in Aristotle and his followers' (*Principles*, no. 11). And when it comes to Being, he says 'the general idea of Being appeareth to me the most abstract and incomprehensible of all' (ibid., no.17).

It could be embarrassing to go further with this, were it not for a seed of truth planted amongst the weeds; and if Berkeley had in fact limited his demonstration to showing that physical substance is from a certain perspective simply consciousness 'congealed', he would not have been too far from the metaphysical reality in play here. To put it in Sir Arthur Eddington's words: 'The stuff of the world is mind-stuff.' *Intellectual substance* would be a better term. Also, if one concedes that the evolutionist position just mentioned is founded on an erroneous hypothesis, then Berkeley is in his full rights in declaring as 'monstrous' the systems erected by 'atheists' to make 'a self-existent, stupid, unthinking substance the root and origin of all beings' (ibid., nos. 92, 93).[1]

He asks 'whether it is possible for us to understand what is meant by the *absolute existence of sensible objects in themselves, or without the mind*. To me it is evident those words mark out either a direct contradiction, or else nothing at all' (ibid., no. 24).

But it is one thing to say that God with His Vision creates the world, and quite another—and ludicrous parody—to put forth that rational man through sensorial perception creates anything whatsoever (apart from working with substances already extant). And holding stubbornly to a sophism of this magnitude, rather than bringing him anywhere within range of true teachings like that of the Advaita Vedânta, lands Berkeley unwittingly nearer a position espoused by the likes of the Church of Christ, Scientist . . ..

The Afghan Sufi philosopher and poet, Jâmî (d.1492), has the last word here:

> Philosophers devoid of reason find
> This world a mere idea of the mind;
> 'Tis an idea—but they fail to see
> The great Idealist who looms behind.[2]

The Scottish philosopher, David Hume (1711–76), considered to be the ultimate skeptic and empiricist with his psychological individualism and atomism, did not delay to turn Berkeley's arguments against their innovator, saying that the Bishop's claim to perception as the source of external reality actually begs the question, since there are no more grounds for believing that mental substance has a foundation than there are grounds for believing the same of material substance; hence *both* must go. Hume adds that we have no way of knowing whether cause and effect even exist; the most we can surmise is based on 'probabilities', but for him this awareness of the limits of rationalism at least brings us to a 'logical positivism' able to serve as a base for scientism.

1. It was likewise Eddington who told the evolutionists: 'If your theory is found to be against the Second Law of Thermodynamics I can give you no hope.'

2. *Lawā'iḥ* ('Flashes of Light'), XXVI; tr. Whinfield and Ḳazvînî, London, Royal Asiatic Society, 1906.

It was this skeptical *tour de force* on the part of Hume that was to awaken Immanuel Kant (1724–1804) from his 'dogmatic slumbers', by the Prussian archirationalist's own admission. To paraphrase the title given Spinoza, Kant might be called the 'pessimist of rationalists'. He knew that Hume had exposed a problem without answering it, and he felt it incumbent on himself to demonstrate that there was indeed a 'more', but a 'more' forever beyond our reach, namely, an a priori *noumenon—das Ding an sich*, the 'thing in itself'—which cannot be directly perceived even though logically presumed and indirectly approached through *synthesizing* 'categories' of thought (whence his term 'transcendentalism' for the 'a priori' science of 'pure reason'). Apposed to this is the 'a posteriori' sense-material known through experience, empirically. For him the gulf between the unknowable *noumenal* content of mind and the knowable or *phenomenal* content is bridged by an 'act of faith', giving the sole 'intuition' possible to the supposedly inaccessible content of a suprasensory order the existence of which we are none the less rationally obliged to admit.

On the social plane, Kant uses the term 'categorical imperative' to define what he again conceives through 'faith' as the duty imposed by free will to act rationally on a moral principle serving as a 'universal law', namely, a sort of ethical teleology.

For us Kantianism is the ultimate disaster of European philosophy; and the basic stumbling block here is the incapacity of a rationalist, claiming intelligence enough to sense something higher, to thus at least allow for the possible existence of an intellectual faculty in fact perfectly well known to countless sages throughout the centuries preceding his little cosmic moment—as Schuon has amply demonstrated elsewhere.

And the irony of it is that the philosopher of Königsberg, in the 'perfect assurance' of his moto *nihil ulterius* (nothing beyond),[1] nevertheless stood on the threshold of the Intellect, for his *noumenon* etymologically derives from the Greek root *noesis*—cognition, intelligence, whence comes *nous*—Intellect. *Noumen* is akin to *nomen*—name, and the *noumena* are none other than the Divine Names or Archetypes, which seen etymologically through the medium of the

1. *Critique of Pure Reason*, p. 395, Max Müller's tr.

Greek verb *phainô*—to show, disclose, bring to light, manifest, shine forth—become the *phenomena* of created existence.

Regarded in this light, Kantianism's role in European philosophy is the ultimate repudiation of the Intellect or primal Subject, thus leaving the way henceforth untrammeled for the exploitation of the outward, material, and basically quantitative pole of existence—an exploitation that is now showing its other face, namely, the *choc en retour* of irremediable cosmic disorder for our world as we know it today.

5 Nominalism to Atomism: The Outer Darkness

> The world destroys Tao, and Tao destroys the world.
>
> *Chuang-tse*, Ch. XVI

SURELY of all the pictorial art from the hand of Albrecht Dürer (1471–1528), the work most enigmatic and disturbing is the copper engraving entitled 'Melencolia'. (Fig. 8) The master craftsman and artist of Nuremberg, living on the threshold of transition between the Middle Ages and our present world, had as stylistic medium at his disposal the late and decadent German Gothic characterized by its exaggeratedly realistic massivity, gnarled forms, and writhing movement. But exceptionally, with the engraving in question here, the medium seems uncannily suited to the subject matter.

Yet what in fact is its subject matter? The first idea that comes to mind is *angst*, a sense of foreboding, the notion of black bile from the original Greek forming the word 'melancholy'; microcosmically or 'alchemically' speaking this is the 'putrefaction' marking the initial stage of the (inner) work which precedes spiritual regeneration and transformation.

The figure that at once strikes the eye in the foreground of this chaotic-seeming tableau is the ponderous, brooding female angel, crowned with a garland of acacia, and seated with a closed book on her lap, while holding idle a pair of compasses. Next to her, in the center of the picture on a rug draped over a stone disk with a cavity suggesting a hub where an axle might go, sits a child angel half asleep; above his wings are the plates of a pair of scales suspended from the façade of a brickwork tower.

The background is a vast body of water surmounted by a rainbow under whose encompassing arch blazes an enormous comet showering sparks throughout the sky. And close to the horizon near the huge tail of the fiery orb soars a monster devil-bat with wings outstretched, the under membrane reading: MELENCOLIA § I.

The interpretation which follows owes much to the article, 'A propos de la "Melencolia"', by Louis Barmont, which appeared in the *Études Traditionnelles* (Paris, 1945–46).

The star of evil augur can portend none other than the 'Flaming Night' that describes the apocalyptic period immediately preceding the dissolution of our present cycle, while the presence of the rainbow is the covenant from Heaven of the dawning of a new cycle, in keeping with the Hermetic interpretation of the initials I. N. R. I. (not in the picture—from *Iesus Nazarenus Rex Iudaeorum*: Jesus of Nazareth, King of the Jews) to read *Igne Natura Renovatur Integra*—By Fire Is Nature Wholly Renewed.

The demon-bat then delineates the reign of Satan, personified in Antichrist, being not only the lowest and most infernal point in our cycle, but also the most transient and unreal, like a shadow sweeping across the cosmos just prior to the intemporal instant that sees the quantitative and qualitative poles reversed, thus inaugurating 'a new heaven and a new earth'. This inversion may be what is symbolized in the double (*yin-yang*) spiral interposed between the strangely spelled word 'melencolia' and the mysterious 'I' (could this latter possibly stand for 'Inversus'?).

The great angel would be the genius, vicegerent, or *shakti* of the cycle that is closing; the large key and four smaller ones attached by a thong to her girdle would then represent respectively the key that opens the cycle, and the keys that subsequently open each of the four ages; the book (of Life) on her lap being shut and the dividers idle would signify that the play of possibilities for our cycle is essentially completed. This idea is corroborated by the presence of a large hourglass suspended from the tower behind her, with the sand running out—the accent on *time* reinforced here by the added presence of a sundial directly above.

The winged infant holds on his lap a writing tablet, and in his right hand a small instrument suggestive of a hammer. Placed as he is in the central position of the picture, under the scales recalling Judgment, and close by a seven-runged ladder reaching out of sight above the cornice on the tower, which ladder can hardly be other than an image of the World Axis—all these elements combined proclaim him to be the genius of the cycle to come.

A greyhound dozes, eyes half opened, at the feet of the angels. Barmont takes this animal to be the mysterious *Veltro* mentioned by Dante in the *Inferno* (I.101–5), who will feed on 'wisdom, and love, and virtue', and who shall chase the beast 'through every city, till he have put her into Hell again; from which envy first set her loose'. The greyhound is referred to as the 'five hundred ten and five' in *Purgatorio* (XXXIII.43), which numbers are generally read to give DVX=leader, who 'sent by God, shall slay the thief, with that giant who sins with her' (the 515 certainly being inferentially set against the 666 in *Revelation* XIII.18).

The juxtaposition of the *Veltro* with the other figures in the engraving would thus be symbolically suggestive of a type of the 'Precursor' to the Second Coming of Christ, in keeping with the Gospel passage: 'Elias truly shall first come, and restore all things' (*Matthew* XVII.11; see also *Malachi* IV.5,6). It is the role of John the Baptist who precedes Christ in 'preparing the way' (*Luke* III.4).[1]

Guénon himself refers to this moment as a 'restoration' (*redressement*) that must happen even visibly before the end of our cycle and the subsequent 'reversal of poles', a rectification to be accomplished by 'he who, uniting in himself the powers of Heaven and Earth, East and West, will manifest outwardly, both in the domain of knowledge and of action, the twofold sacerdotal and royal power conserved across the ages, in the integrity of its unique principle, by the hidden custodians of the Primordial Tradition.'[2] The French metaphysician adds that it would be useless to seek to know in advance when and how this could come about, all that one might imagine on the subject almost certainly differing from the eventual reality.

St. John the Baptist, whose feast day is June 24, is identified with the summer solstice as the Saint John of Summer, just as St. John the Evangelist, whose feast day is December 27, is called the Saint John of Winter, thus being identified with the winter solstice and the time of the Nativity, marking the affinity between Christ and this disciple

1. It is noteworthy that in Eastern iconography, St. Christopher (*Christophoros* in Greek meaning 'one who carries Christ') is often represented with the head of a dog as a mark of fidelity.

2. *Aperçus sur l'Initiation*, Paris, Éditions Traditionnelles, 1946, p. 264.

'whom Jesus loved'. We are speaking here of a complementary polarity or affinity between the functions of the two St. Johns.

But the Forerunner to Christ, in his association with the summer solstice, is thus in some manner allied with the constellation of *Canis Major*, due to its heliacal position at this period. Which brings us back to the *greyhound* in the engraving as a figure of the Precursor who will pave the way with wisdom, love, justice, and virtue for the coming of Christ cyclically manifesting at the darkest point of the year, namely, the instant a New Year commences. Analogous allusions to this preparatory function are to be found in Hinduism (the Kalki Avatar), Buddhism (the Maitreya Buddha), and Islam (the Mahdi).

Coming back to the 'Melencolia', at the foot of the stone platform supporting the great angel are Dürer's initials, and the date 1514. The same figure appears in a magic square of sixteen numbers that is set into the tower over the angel's wings, and with a bell above. The sequence of these numbers from whichever direction added gives the sum of thirty-four, which Barmont relates to matters concerning the Templars, but of a nature too conjectural to include here. Other elements in the picture are a crystal sphere in front of the dog, and at the foot of the ladder an immense stone dissymmetrically hewn—filling the space as an unwieldy quantitative mass. In striking contrast is a snug castle on the banks of the shore, distantly framed between two rungs of the *axial* ladder at the level of the infant's head, where the rainbow begins. There is something suggestive here of the Celestial Jerusalem (*Rev*.XXI).

Further objects lying about are the tools of carpenters and masons. The engraver was possibly indicating the artisanal source of initiation for the Hermetical science at his disposal, or what is known as the 'Lesser Mysteries'. Although Barmont does not mention it, there is a striking axial or pivotal geometry in the composition of the picture, based on a cruciform theme. The horizontal element is massive and *quantitative*, in a line running from the preponderantly substantial angel, through the dissymetrically hewn stone block, to the devil-bat and the comet's tail. The vertical element, relatively more delicate and *qualitative*, is composed of the crystal sphere, the greyhound, the child-angel, castle, scales, ladder and rainbow. It is the hub of the stone disk (the future cycle of existence?) that forms the focal point of the cruciform theme, which point metaphysically is where each state of existence is traversed by the vertical Axis.

Nowhere is there visible any 'Christian' imagery: Dürer like Shakespeare (and the alchemists) knew to keep his symbolism purely 'pagan'. One sees what it was at this 'liberated' time to be an imprudent Christian esoterist, with a figure like the Italian Dominican and somewhat enigmatic Neoplatonic Hermeticist, Giordano Bruno (*c*1548–1600), being burned at the stake.

★

Dürer was thus giving witness in his way to the fact that the 'brave new world' just over the cosmic horizon was going to be of a different cast from what sanguine Renaissance Man and his immediate posterity anticipated. And there were others who would share Dürer's view.

Only fifteen years after the death of the German engraver did the heliocentric system of Copernicus, his contemporary, first appear in print—the same year (1543) its author died. But it was reserved for Galileo concretely through the aid of the telescope to open new *outward* vistas onto the vastness of the physical cosmos, thus considerably shrinking concepts hitherto held of the Earth's centrality, size, and importance. His treatise *Sidereus Nuncius* (The Starry Messenger), appearing in 1610, brought him acclaim throughout Europe as the founder of descriptive astronomy. This was twenty-three years before his ill-starred collision with ecclesiastical authorities, understandably alarmed over the secular implications of a system supplanting the celestial structure to which the Church adhered.

Actually, on a purely cosmographic level, there is no really stupendous distinction between a geocentric and a heliocentric view. In the words of Daniel Boorstin, 'Copernicus would not change the shape of the system, he simply changed the location of the bodies'.[1] And this in turn is relativized once one realizes, in the exposition of Wolfgang Smith, a contemporary physicist, 'that twentieth-century physics is in fact neutral on the entire issue. . .. What modern physics insists upon—ever since Einstein recognized the full implication of the Michelson-Morley experiment[2]—is that the concepts of rest and

1. Daniel J. Boorstin, *The Discoverers*, New York, Random House, 1983, p. 295.

2. An attempt made in 1887, and subsequently repeated, to measure by means of an interferometer the fluctuations in velocity that the speed of light should theoretically exhibit when calibrated from different angles against what should be the earth's orbital motion. All efforts failed to detect any appreciable variations whatsoever in the presumed medium of a 'luminiferous ether', or the presence of an 'ether wind' (our note).

Figure 8

'Melencolia', by Dürer.

Figure 9

Ceramic tilework, French, 16th-17th c.

Figure 10

Passage from the world of pairs to the world of unity. Four Lohans and Two Attendents Witnessing the Transfiguration of a Lohan.

an Aristotelian, while accepting much in Copernicus, was yet unwilling simply to abandon Ptolemy; hence he sought a *via media* between the two systems by retaining an immobile Earth but allowing for the five planets to circle the Sun, which together with its planetary cortège was considered by him to revolve about the Earth.

Even Johannes Kepler (1571–1630), regarded as the founder of physical astronomy, started out both as a theologian and an astrologer who believed in the 'harmony of the spheres' and interested himself in the 'musical intervals' of planetary movements. But he eventually arrived at the conclusion that celestial physics would make more sense if the motivating power was understood to be not an *anima* (soul) but rather a *vis* (force). His 'contribution' is thus to have 'succeeded' in eliminating the last vestiges of a vertical perspective: in the words of Boorstin, 'he boldly pointed the way from an organic to a mechanical explanation of the universe. "Spirits" and "celestial intelligences" would be replaced by forces'.[1]

This made for the entrance of Sir Isaac Newton (1642–1727) with his gravitational forces operating a 'clockwork' universe put into motion by God, being in the English mathematician's conception 'a natural cause, very well skilled in mechanics and geometry'.[2] We now have on a modern scientific footing a *fully exteriorized* physical world *to which* but not *in which* God is present,[3] a world that lends itself to 'experimental philosophy', as Newton puts it in his *Principia*.

We said earlier that no really stupendous distinction exists *cosmographically* between the geo- and heliocentric positions. From the point of view of this paper, the Earth as Center relates more strictly to the pole *Subject* or the Infinite, and the Sun as Center more to the pole

1. Boorstin, op. cit., p. 311. It can be seen that we are fast going counter-evolutionistwise from higher to lower forms. It is noteworthy that Harvey and Boyle at this period were carrying medical concepts of the human body in a mechanistic direction.

2. Newton in a letter to Dr. Bentley, 1692; see *Selected Mystical Writings of William Law*, ed. by Stephen Hobhouse, London, Rockliff, 1938–49, p. 418. Newton also said in his *Opticks*: 'God in the beginning formed matter in solid, massy, hard, impenetrable, movable particles, of such sizes and figures, and with such other properties, and in such proportion to space, as most conduced to the end for which he formed them.' And this is how man perceives the world today.

3. This nuance appears in Newton's *Opticks*, Query 28; Hobhouse, op. cit., p. 417—a nuance confirming the statement in our Preface 'on the premises of a Newtonian universe'.

Object or the Absolute. Hence, to say that the Sun is the center of our *solar* system is a little like saying God is the center of our universe: this is—or ought to be—obvious beyond need of expression.

Yet *experientially*, our world *is* the center. We still use perforce a geocentric system for navigation and other matters of orientation, including astronomical measurements. And on this earth each ego as *subject* consciously or unconsciously feels itself existentially to be central. In this perspective the Ptolemaic system of astronomy comes closer than the Copernican to recognizing the complementarism subject-object. The reason for this is that in any manifested universe where one speaks of a subject, it goes without saying that the very term 'manifestation' obviously implies an object. But the inverse is not so easily seen to be true; and in a heliocentric cosmographic system where the physical object is perceived as the whole of reality, the cognizant subject is inevitably going to be relegated sooner or later to a status little more than that of 'onlooker', or sort of cosmic 'pariah'—a situation ever more in evidence today.

The fact is that in terms of polarities, heliocentric/geocentric are inseparable corollaries. Hence in some mysterious fashion reaching cosmologically beyond the powers of our earthbound faculties to seize the intangible and universal measure and reality of the matter, both concepts are valid and interdependent. The *Qur'ân* concisely situates this mystery: 'It is not for the sun to overtake the moon, nor doth the night outstrip the day. They float each in an orbit' (XXXVI.40).

It pleased Neoplatonists like Nicholas of Cusa (1401–64), before the time of Copernicus, and Bruno, in the wake of the Polish astronomer, to espouse for reasons of metaphysical aptness the heliocentric view, since in relegating our world to the peripheral pole it opened vistas Oriental in magnitude onto the plurality of worlds and cycles possible in a universe vast beyond conceivable measure (the *Samsara*).

But the common run of humanity is not constituted for metaphysical insight, and the same perspective that expands the soul of a 'pneumatic' tends to shrink the vision of the 'hylic' and the 'psychic'.

In a Ptolemaic world, man inherits a sense of *noblesse oblige* based on the concept of centrality; his place in the hierarchy of human possibilities falls naturally to his appointed portion in an integrated 'caste system', ranging from spiritual authority and temporal power, to the mercantile and artisan vocations, followed by the husbandman's calling, and lastly the sustaining laborer. The world for traditional mankind is the manifested center that reflects the workings of the Center *in Divinis*; and the Sun that lights and vivifies the physical universe is regarded as the outward form of the inward and Supernal Sun that illumines all the worlds.

In a Copernican setting by contrast, man *de facto* is pushed aside to the periphery as a speck on a cinder exploded from a 'furnace' (this is what they call it) in the sky that is just one of many million or maybe billion or trillion more 'furnaces'.

To reduce these radically schematic formulations to a purely axiomatic opportuneness, one could say that the normally 'vertical' outlook of the geocentric makes him spiritually a 'heliocentric', whereas the typical heliocentric by the materialism of his 'horizontal' outlook is to all intents and purposes a 'geocentric', i.e., 'earthbound'.

★

That others would share the somberness of Dürer's view is evidenced in the work of a poet like Edmund Spenser (1552?–99):

> Me seemes the world is runne quite out of square,
> From the first point of his appointed source,
> And being once amisse growes daily wourse and wourse.
> . . .
> Right now is wrong, and wrong that was is right,
> As all things else in time are chaunged quight,
> Ne wonder; for the heauens reuolution
> Is wandred farre from where it first was pight,
> And so doe make contrarie constitution
> Of all this lower world, towards his dissolution.
>
> (*Faerie Queene*, Prologe Bk. V)

And in the work of another poet like John Donne (1572?–1631), who had even taken the initiative to meet Kepler, and who would not wish to challenge Copernicanism, while yet fearing the baneful influence heliocentrism could have on 'man's mind':

> Then, as mankinde, so is the worlds whole frame
> Quite out of joynt, almost created lame.
> . . .
> And new Philosophy calls all in doubt,

The Element of fire is quite put out;
The Sun is lost,[1] and th'earth, and no mans wit
Can well direct him where to looke for it.
And freely men confesse that this world's spent,
When in the Planets, and the Firmament
They seeke so many new;[2] then see that this
Is crumbled out againe to his Atomies,
'Tis all in peeces, all cohaerence gone.
. . .
For of Meridians, and Parallels,
Man hath weav'd out a net, and this net throwne
Upon the Heavens, and now they are his owne.
Loth to goe up the hill, or labour thus
To goe to heaven, we make heaven come to us.

(*An Anatomie of the World*)

The same as with Donne seeking out Kepler in Austria, John Milton (1608–74) on a trip to Italy in 1638 called on Galileo, then blind and decrepit and in the poet's words 'a prisoner to the Inquisition'. He considered the astronomer to be the victim of a just cause, and yet—as Boorstin has remarked—when Milton two decades later wrote *Paradise Lost*, he saw himself obliged by the nature of things to base his poem within a Ptolemaic perspective.

*

The message we are trying to get across is that people are entranced with a physical world by its inherent structure seemingly independent and solid, which in fact is anything but that. One cannot have it both ways: either the universe is an organic whole, a *uni-versus* or 'one-turning', or else it is an organic/inorganic chaos, being a multiplicity both of combined and disconnected physical particles fragmented throughout a space reckoned in light-years of immensity—billions of them. One cannot, that is, claim for our solar system the concept of ordered and *measurable* subject-object coordinates which once outside

1. In the sense that when seen as encircling (encompassing) the world, it 'contains' the world, as analogously, according to their various 'zodiacal' properties, do the Moon and planets likewise contain the world, being the celestial instruments guiding its operations – a harmony which Donne considers as now disrupted.

2. A reference to Galileo's revolutionary telescopic sightings. The line following suggests a cosmic dispersion crumbling into atoms.

the solar system somehow *cease* to function at some undetermined point, or 'heliopause', being replaced by purely physical explosions and contractions playing out a chaotic autonomy—as if such were possible within the inexorable principle of heteronomous polarity that determines all created things.

Guided by the same propensity for spiritual opportuneness that inspired the mediaeval cartographer with an essential and not just accidental, however physically accurate, frame of reference, the mediaeval astrologer/astronomer (these two disciplines then constituting a single science) likewise based his view of the heavens on the intrinsic and not merely phenomenal reality of things. Earth as object and living crown jewel of the solar system—and not *just* a purely 'physical' planet like the others—was thus *contained* at the center within a series of concentric spheres, beginning with the Moon's orbit, followed by Mercury, then Venus, and next the Sun in the median—hence central—position, with Mars coming afterwards, then Jupiter, and finally Saturn, the planetary sphere immediately preceding the Firmament of fixed stars. It is the apparent West-East displacement of this Firmament along the ecliptic that lets us gauge the precession of the equinoxes by which the Platonic year is measured. As regards the trans-Saturnian planets Uranus, Neptune, and Pluto, they can be considered in one way or another as incorporating something 'extra-territorial' in their nature, and have, moreover, attained cosmic prominence only recently, in concert with the world's increasing instability. Ultimately comes the ninth circle, this being the Empyrean or starless heaven where time and Eternity mysteriously conjoin (as with the mystery of the split-gate symbolism broached in Ch. 2).

In this concept of Uni-versus, it is the Object that is *contained* by the Subject, or the Known by the Knower, in keeping with the dictum of Plotinus (205–70 AD): 'Soul (*Nous*) is not in the universe, on the contrary, the universe is in Soul' (*Enneads*.V.v.9). This idea tallies with the teaching of Erigena and others, that Man is a microcosm or abridgement of the Universe; and with the statement in the *Upanishads*, that '*Âtman*, which dwells in the heart . . . is greater than all the worlds together' (e.g. *Chandogya Upanishad*, III.XIV.3,4). The Greek philosopher Parmenides (fifth century BC) situates the rationale for this in saying, 'nor is being such that there may ever be more than what is in one part and less in another, since the whole is inviolate.'[1] The Zen Buddhist abbot Takuan (1573–1645) gives the same perspective a Far-Eastern coloring: 'When mind is nowhere it is everywhere. When it occupies one tenth, it is absent in the other nine tenths'.[2]

From the time of Galileo onwards the 'one tenth' outlook corresponds more and more to man's evolving position, his erstwhile spiritual and qualitative stature ever shrinking with the expanding physical view of each more powerful telescope, while simultaneously his technology burgeons spectacularly due to the central position this late-cycle man accords the outward, objective, and preponderantly material and quantitative pole of existence.

There is nothing inherently wrong about the 'out there' perspective so long as it is contained as an extension of the 'in here' perspective, but in practice with the degeneration of humankind this balance proves almost impossible to realize—a point brought home by Plato concerning the downfall of Atlantis (*Critias*, 121). The 'out there' view ever seeks further and further autonomy. Applied science when seen in the light of the whole Essence-Substance schema offers a technology utilitarian at best, trivial at the least, and psychically—hence by extension environmentally—corrosive at the worst; but the results of these 'out there' manipulations are so overwhelmingly beguiling that humanity is constantly drawn deeper and more deeply still into this cosmic 'conspiracy'.

And if a major road-block looms ahead, the answer always is: 'Science will fix everything'. But when the inevitable Day of Reckoning arrives, the situation will be tantamount to that in which the disbelievers and idolaters described in the *Qur'ân* find themselves when calling *in extremis* for the help of Satan on whom they had relied, and whose searingly cynical rejoinder is:

> Lo! Allah promised you a promise of truth; and I promised you, then failed you. And I had no power over you save that I called unto you and ye obeyed me. So blame me not, but blame yourselves. I cannot help you, nor can ye help me. Lo! I disbelieved in that which ye before ascribed to me (XIV.22).

1. *From Thales to Plato*, ed. T.V. Smith, Chicago, Univ. Chicago Press, 1934 and 1956. This passage anticipates Bell's theorem, quoted later in the chapter.

1. D.T. Suzuki: *Zen and Japanese Culture*, London, Routledge & Kegan Paul, 1959, p. 108.

And the exquisite incongruity of man's being ever more engrossed with the fruits of his manipulations of the outward environment is that he as *faber* is simply exteriorizing narcissisticwise his own subjectivity! Every last jot, whit, and tittle of the technology elaborated for man's delectation issues from the subjective pole. The inorganic (so-called) universe has not a finger to raise on its own behalf—even though it will not be left unanswered. For the modifications man brings to the outward environment, *his is the entire responsibility*. If his soul is deiform, the modifications will reflect this. If his soul is godless, the modifications will reflect this. Hence the ancient Egyptian inscription on a scarab placed over a mummy's breast: 'O my heart, rise not as a witness against me'.[1]

The objective pole in other words can never be anything except a mirror of the subjective, since it is this latter pole that contains *all* the initiative. If its directives are angelic, the image will be angelic; if its directives are Titanic and Promethean, so accordingly will be the image, and so forth. In our post-religious age, the objective pole is made to spew fort the detritus of the individualism, relativism, and aberrancy which together result from the divorce of the subjective pole with first principles and with human norms as established from time immemorial. Having lost contact with the Source of its own inwardness—the Primordial State—the subjective pole gravitates towards the outward dimension and seeks its paradise in the bewildering panoply of phenomenalism—from the frivolous to the frankly sinister. In a word, the change from the Ptolemaic to the Copernican perspective has altered not one iota what we receive from the universe, but it has altered radically what we give back to the functionings of the universe within our reach.

A shift from geo- to heliocentric does not affect the relationships of the movements. Zodiacal positions, for example, remain unchanged. And even though in our solar system the Sun is central and orbited by the Earth along with the other planets, *the Earth alone* in this system is served (Zodiacally) by the Sun and planets, and not vice versa. That the Earth goes around the Sun is something any schoolboy knows. What the schoolboy does not know is the marvelous harmony and intricacy of the entire celestial system as understood by the ancients. Copernicanism adds nothing extraordinary, whereas something irre-

1. In James H. Breasted, *A History of Egypt*, 1906.

trievable is lost with the rejection of Ptolemaism, which served as a symbolical 'chart', in contradistinction to scientific 'fact'. What the schoolboy is not taught is that Ptolemy was writing, not from an age of ignorance, but from an age of priorities other than ours, where the vertical and spiritual perspective of Platonic (similar to Vedântic) emanationism was given preference to a horizontal and material vision such as ours based on purely physical facts—'facts' which become less factual in their own turn with the advance of quantum physics.

For science, despite all its specialized brilliance, has not come that near to revealing the ultimate workings of the outside universe. Even recent studies of the rings of Saturn—which are less than light-years away—have raised more problems than answers. To compensate for their innate sense of inadequacy, the present generation of astronomers (and physicists) have exploited a 'jocular' vocabulary (to the end of reinforcing their egos) that belittles cosmic immensity (and minuteness): terms like 'big bang', 'big crunch', 'dark holes', 'worm holes', 'quarks', MACHOs ('massive compact halo objects'), WIMPs ('weakly interactive massive particles') and the rest. 'Heliocentrics', in other words, are not for their 'advanced learning' all that more objective.

The truth is that as beings in time and space our 'subjectivity' is more important to us than our 'objectivity' qua *objects*: as object the tendency is to be dispersed into indefinity; it is our subjectivity that prevents this. Although modern man *theoretically* considers the world as being no more than a speck of dust hurtled through the universe, in *practice* he treats his particular space and time as central. Temporally speaking, even the past of no more than fifty years ago is now viewed by him with a certain cavalier condescension as being in rapport to the present moment, relatively 'naïve', 'simplistic', 'undeveloped'; the medicine of 100 years ago is now considered to have been positively 'primitive'. It is this attitude that someone in speaking against evolutionism has rightly dubbed as 'chronological snobbery'; what is certain is that without this attitude there never could have been a Theory of Evolution.

★

We are now in a better position for understanding how contemporary man can feel comfortable with evolutionism—a question raised at the conclusion of Chapter 1.

This capacity for gravitating towards the irrational seen as rational is the result of man's loss of contact with his inward center, combined with his innate sense of centrality notwithstanding. This, plus centuries of conditioning by bad philosophical systems, culminating in the dialectical materialism of Karl Marx (1818–83) and Friedrich Engels (1820–95), a construction which holds that the observable world is real without any transcendent origin, a world independent of the mind of man, since matter by their perspective is logically and temporally prior to mind, itself being judged as nothing but an outgrowth of this matter.

The setting is thus ripe for the fateful entrance of two of their contemporaries: Charles Darwin (1809–82, whose *The Origin of Species* was of seminal importance to Marx's *Das Kapital*) and Herbert Spencer (1820–1903)—the former to expound the specifics of 'organic evolution', and the latter to fill in the moral and philosophical overtones. No matter if much of their theory had already been anticipated by the French naturalist Jean Baptiste Lamarck (1774–1829); the credit went to the other side of the Channel.

Reaching into the unknown depths of the Outer Darkness, Darwin with his inherited background of bad European philosophy could thus 'materialize' an amazing progression of 'upward' emanations, beguilingly flattering to a contemporary humanity now supplied with all the arguments necessary for unabashedly presuming itself superior even to Caesar and Christ by the circumstance alone of chronological succession. In the opacity of their nineteenth-century vision, Darwin's followers elaborated a portrait of the 'grossness' of early man; perish the evidence that this primitive creature none the less believed in a transcendent and cyclically atemporal origin the opposite of evolutionism.

To repeat: we have by now sufficiently pursued the genesis both philosophical and existential of the widening breach over the centuries between subject and object, to better grasp how it has come about that the object 'out there' has finally attained the status of a 'totally independent physical entity', in the dictionary definition given at the beginning of the first chapter.

Earlier we said that Christianity is essentially a sedentarily fixed religion. In mediaeval Christendom the gravitational center was the bishopric with its cathedral being a shared sanctuary uniting the spiritual, intellectual, sovereign, mercantile, artisanal, and agricultural activity for each sector of humanity concerned.

Coming down now to Darwin's world and century, domestically the quantitative object or machine was implacably replacing the qualitative craftsman or subject. Abroad, the same sedentarily situated man was busy violating his traditional Christian bonds and setting his sights on the farthest stretches of our globe in a work of colonizing and exploiting for material ends those materialistically unprepared peoples who by the same token were still too inwardly oriented or 'primitive' to gauge correctly the intangibles of the encounter. And astronomically, man was setting his telescopic sights on the farthest reaches of the physical universe in a work of divining his very origins.

Modern humanity was in other words by this time virtually conditioned to believe that the ultimate answers to the 'riddle' (so called) of existence were to be found 'out there' somewhere. The problem is, the 'out there' cannot exist without its Source, the 'in here'. The objective world can be extended as much as you like away from the 'in here', to the point that the 'in here' is no longer perceived or believed in. Yet invisibly perforce it has to continue to function, at the risk of the world's extinction. For stretched beyond a certain point, the link between inner and outer is sundered, which should that happen, the objective or phenomenal world would be volatilized, being instantaneously reduced to its primordial or essential substance, so that the world we know would be gone forever.

★

When the 'Titanic' on the eve of the First World War struck an iceberg and sank (15 April 1912), it was more than just the passing of that world's mightiest ship; it represented beyond that the passing of an era—the apogee of the earth's greatest solidification and 'hardness'. More and more now would there be ascending from below and cracking through the comfortable material carapace of Newton's 'clockwork' universe the corrosive influence of relativism, which replaces immutable Being with mutable becoming, as witnessed for example in the work of Sigmund Freud (1856–1939) and of Albert Einstein (1879–1955). Modern man, having virtually lost all concept of the Absolute, was henceforth passively letting himself be progressively locked into an 'absolute' relativity from both sides: subjectively through the theories of the Austrian neurologist, and objectively through those of the German-American physicist.

It is not that Einstein's *theory* of relativity is within certain relative contexts altogether fallacious, being as it is a play of polarities, but that our universe—relative by definition—has always adequately symbolized a hierarchy of integral order giving onto the 'relatively absolute'[1] immutability of the supraformal status of Being: whereas Einstein's theory 'absolutizes'[2] the relative, thus in its turn (other considerations aside) placing the pole *material object* despite everything on the pedestal of autonomous entity, hence making his hypothesis effectively an appendage of evolutionism.

Freud for his part—as supereminently with his fractious alter-ego Carl Jung (1875–1961) later—did not, in plumbing the depths of the subconscious, come up with mere chimera (which would have been a matter of little importance); instead, his work has opened irreversible access to the subliminal world of influences from the lower and eminently subjective and relativistic prolongations of the psychic domain or inferior reaches of the subtle order, which border on what is truly infernal. The pattern is really an inversion of the way in which primordial man saw the Archetypes through the transparency of things. With Jung in particular we are back to 'archetypes' again, but from below rather than from above, from the nether realms rather than the celestial. The basic criterion here for those who have the eyes to see, combined with a sense for the Sacred, is the character at once of parody and dissociation which distinguishes these fragmented, murky, and subrealistic forms from true Essences.

But the point being made at present is that the subjective pole is again gaining prominence in a world which has cyclically passed beyond its materialistic peak of greatest objectification or exteriorization, only, this time the inward focus is ever more 'downwards' and ever less 'upwards'. The role of evolutionism in this process is absolutely indispensable for the 'Powers of Darkness': to keep man marching in the illusion that this downwardness is all part of a progress 'upwards' towards the realization of greater and greater potentialities—right up until what will ultimately emerge as the day of the Great Profanation that immediately precedes the 'reversal of the poles', as explained in Chapter 3. This final inversion, be it said, by its nature necessarily escapes anything the negative powers in question could in the least imagine or suspect on the subject.

1. The expression is Schuon's.

2. For him the 'absolute' by which all else is to be measured is the speed of light, conventionally given as some 186,000 miles per second.

The whole matter of the rapports between subjectivity and inwardness is actually very simple; for it goes of itself that if the Kingdom of Heaven is within, so equally by the law of polarities is the kingdom of hell (modern sensibilities notwithstanding); and this is precisely why we have religions—to see that our journey inwards is upwards and not downwards, a spiritual recollection and repossession, and not a psychic dissociation and possession leading to irremediable ruin. It should be emphasized that the role of deviated cults in all this is merely that of an inner door to an outer darkness.

Now with the passing of a predominantly 'physical' materialism there is yet another aspect of inwardness that has to be taken into account, and which is particularly microcosmic, namely, the quantum world of the infinitesimally small, engendering theories about the working structure of what we call 'matter' which are replacing Newtonian concepts. The abhorrence of the great English physicist for the ideas to come manifests in his presentiment:

> That one body may act upon another at a distance through a vacuum without the mediation of anything else . . . is to me so great an absurdity, that I believe no man, who has in philosophical matters a competent faculty for thinking, can ever fall into.[1]

Einstein was probably the last major physicist to believe in the unqualified autonomy of matter: 'The belief in an external world independent of the perceiving subject is the basis of all natural science.'[2]

The incongruity in this remark is that his own relativity theory is observer-dependent, once given that space and time are relative to a perceiving subject's constant displacement in those dimensions.

Whatever the case, the 'inwardness' of microcosmically oriented quantum physics has to do with the 'way in *there*' (not in *here*) of subatomic entities, being thus the counterpole to the 'way out there' of macrocosmically oriented astrophysics; but with the difference that the physicists by the very structure of their findings, unlike the astronomers, know that they are dealing not with the nature of essential reality but simply with abstract mathematical phenomena, with processes and equations concerning the invisibly infinitesimal—

1. Cited by Nick Herbert, *Quantum Reality*, New York, Anchor Press/Doubleday, 1985, p. 213. 2. Idem, p. 201.

no matter what may be the intimate constitution of electronic waves or particles.

Now some schools at least, in the new physics, break with Einstein's belief in an external world independent, without qualifications, in that they throw in a hefty qualifier to the effect that reality is in some manner observer-created, no less (the school of John Wheeler), or further, that *consciousness creates reality* (the school of John von Neumann). This to be sure is not a complete concession to Bishop Berkeley's *Principles*, since the physicists in question do not for an instant doubt the absolute existence of what is popularly called 'matter'. But where they approach Berkeley is in their laboratory-based experiments indicating that the dynamic attributes of this 'matter' depend on human observation for their functioning.

★

A further quantum leap was taken in the mid-sixties by the British physicist John Stewart Bell, during a sabbatical leave from the CERN accelerator center in Geneva. At a time of renewed cyclical instability when spiritual authority was collapsing under the modernist-orchestrated assault by Vatican II, and juridical authority collapsing under the permissiveness of the Warren Court, and social authority the world over losing its moral bearings, the British scientist came up with an observation literally galactic in scope, eventually to be known as Bell's Interconnectedness Theorem. By way of laboratory experiments with photons of green and blue light polarized at different angles through green and blue calcite crystals, and the signals that ensue, plus a bit of mathematics, Bell was able to demonstrate something quantum physics has yet to disprove, namely, that *the nature of reality has to be non-local.*[1]

What this purports to mean is that between our world and even the most distant galaxy imaginable there is a superluminal instantaneity exempt even from the speed of light, a pervading reality that defies all concept of definitive localization. What this further means is that the stuff of reality is essentially independent of time and space. And this in turn means that the field of quantum theory is grazing the threshold of the subtle order, which itself is free from all spatial conditions, as well as from the characteristic limitations of the temporal.

But one could hardly compare this finding with a descent of angels

1. This material is taken from Herbert, op. cit.

bringing glad tidings from on high, for here we are dealing with the underbelly of matter, groping our entrance, as it were, into the subtle sphere, not by way of its prototypal nature as the immediate principle informing the corporeal world, but rather by way of its 'atomic' or dissolving capacity through the fissures that arise in the carapace of materiality once its point of greatest solidification, density, or hardness has been passed—a setting which lends itself supereminently to the abstractions of quantum reality.

Yet taken in itself without reference to its modernist context, not only is Bell's theorem perfectly neutral, corroborating moreover what has always been known in the domains of spiritual and cosmological sciences, it is above all fascinating for the insights it imposes on those who are conditioned to respect the data of modern science but not the teachings perennially transmitted by traditional sources.

As an example, oceanographers working on the behavioral ecology of the hammerhead shark, in the Gulf of California, find the creature's navigational precision uncanny in its long direct voyages to feeding grounds and back, and suppose it due to sensors on the shark's snout that pick up Earth's magnetic and electric fields to give the predator its bearings.

The marine biologists could spare themselves these labored conjectures simply by invoking Bell's theorem, namely, that a non-local connection identifies the sharks with the feeding grounds, in such wise that they have but to swim whatever the distance required, to where they 'already are'. The same principle can be applied to the unerring sense of direction exhibited in the distant migrations of birds and fish, and insects as well.

Now Bell's discovery has to do with what he calls the nature of *reality*, for it is obvious that non-localization cannot be posited of manifested objects in themselves, at the forfeit of one form being indistinguishable from another. Non-localization is the property of what the Hindus call *âkâśa* ('ether'), and what in Western traditions—Platonic, Hermetic, Kabbalist, and Alchemical—is variously known also as the Quintessence, World Soul (*Anima Mundi*), or Primary Matter (*Materia Prima*), it being both the subtle prototype and substratum of our physical universe.[1] The instinct of animals has in

1. See on this subject my article, 'The Quintessential Nought', *Studies in Comparative Religion*, Winter 1977. Thus Bell, while not naming it, effectively brings back the ether which Einstein had to abolish in order to make way for his special relativity theory.

some manner to be linked with this for their guidance; it is pre-eminently the world in which the shaman functions—the world of operative magic as distinguished from mere legerdemain. It could well be contact with this domain that aided the ancient Polynesians in their Pacific migrations. And that these people still know the art of navigation through simultaneous contact with the physical, subtle, and spiritual domains has been fully attested to.[1] They learn to 'see with the heart' and 'breathe with the feet'.

Through recourse to what had to be the subtle order the jinn were seen able to transport the likeness of the Queen of Sheba's throne to Solomon's palace 'within the twinkling of an eye' (*Qur'ân*, XXVII.40).

Says Hermes: 'The incorporeal cannot be enclosed by anything; but it can itself enclose all things; it is the *quickest* of all things, and the mightiest.'[2] In the same vein, Eugen Herrigel cites the Zen master archer's pithy definition of attainment in his art: 'Inside and outside are united in this . . . the archer hits the target without having aimed—more I cannot say.'[3]

Astrology in turn might be less of a bugbear to modern astronomy if it was understood in the light of Bell's theorem that it is not a question of planets having some kind of physical influence—gravitational or otherwise—on terrestrial configurations and events, but rather that the planets—each one with its particular property—operate here like all other celestial bodies, from the 'sphere' of non-localization, and thus independently of relative volume and distance. The Zodiac, in other words, is created 'inside' us: 'In the Soul is the spiritual circle or Zodiac, wherein is fulfilled what the twelve signs represent in the natural Zodiacal system of the Universe or Macrocosm.'[4]

Bell's theorem[5] can also throw light on the matter brought up in

1. (See James Barr: 'Of Metaphysics and Polynesian Navigation', *Avaloka*, Vol. III. nos. 1 & 2.

2. Scott, *Hermetica*, *Lib*. XI(ii).18 (italics ours).

3. Eugen Herrigel, *Zen in the Art of Archery*, London, Routledge & Kegan Paul, 1953, p.80.

4. John Ward (1781–1837): *Zion's Works: new light on the Bible from the coming of Shiloh, the spirit of truth*; Edinburgh, 16 Vol., 1828–37; II. p. 238. See Paracelsus: 'The starry vault imprints itself on the inner heaven of a man' (*Paragranum*: in Harris L. Coulter, *Divided Legacy*, Washington D.C., 1975, Wehawken Book Co., Vol. I, p. 398).

5. Actually, his discovery is corroborated by other findings in quantum mechanics not directly related to it, findings called 'counterintuitive' evidences. Huston Smith gives us some examples from these never-never realms in his *Forgotten Truth: The*

Chapter 4 of a cult still famous in the Far East centering on a Buddha living 43 billion, 200 million solar years ago. If the theory of interconnectedness means what it says, that an invisible reality which is instantaneous, unmediated, and unmitigated obtains throughout our cosmos, claimed by astronomers to be spatially at least 26 billion *light years* in diameter, then what after all by temporal comparison are a 'paltry' 43 billion *solar years*? And this for the rest makes St. Nicholas positively contemporary, as he still is moreover in the Eastern and some sectors of the Western Church.

All of which is a way simply of saying that the closer one comes to the internal and spiritual heart of things, the less do external historicity and temporality matter. It is the 'nowever' and atemporal cosmic cyclicity that Mircea Eliade (as already mentioned in Ch.4) has named the 'Eternal Return' (ἀυακύκλησις = circuit, revolution, cycle), which was the world view predominant up until the time when the Renaissance replaced it with the historical, temporal, and secular perspective.

Cyclical orientation is archetypal, vertical, static, and essential; temporal orientation is linear, dynamic, horizontal, and accidental—historicity being an affair of the ephemeral and transient. Naturally the two perspectives cannot be entirely dissociated one from the other: it is a question of where the emphasis falls.[1]

As our era advances towards the confines of the quantitative and objective pole, cosmic disequilibrium cannot but increase, given that even the most negative possibilities of manifestation must have their day ('for it must needs be that offences come . . .'). But the qualitative and subjective pole, though progressively hidden, is ever present; and the upward current towards this pole is (in a 'yin-yang' sense), for

---

Primordial Tradition (New York, Harper & Row, 1976), pp. 105–6: 'Parallel lines meet, curves get you from star to star more quickly than do Euclid's straight lines, a particle will pass through alternative apertures simultaneously without dividing, time shrinks and expands, electrons . . . jump orbit without traversing the intervening distance, and particles fired in opposite directions, each at a speed approximating that of light, separate from each other no faster than the speed of light.' We are a long way from Ptolemy, for once a unified world picture is shattered, there is no second unity to replace it with.

1. These perspectives have been expounded more fully in the paper already mentioned in Ch.3, 'The Dragon that Swallowed St. George'.

those who can find it, all the stronger for flowing against the cyclical tide.

★

A Moroccan holy man told us not so many years ago that 'from now until the end of time, all the doors are wide open—the gates to Heaven and the gates to Hell.' These words if they are to be believed say much about our present moment cyclically; and they afford a telling insight into what the mysterious and disquieting tableau by the master engraver of Nuremberg is all about.

6 The Sundoor

Two birds, fast bound companions,
Clasp close the self-same tree.
Of these two, the one eats sweet fruit;
The other looks on without eating. . . .

When a seer sees the brilliant
Maker, Lord, Person, the Brahma-source,
Then, being a knower, shaking off good and evil,
Stainless, he attains supreme identity (*sâmya*).

Muṇḍaka Upanishad, III.i.1,3[1]

OVER the fireplace in the lounge of the Hôtel du Grand Monarque at Chartres is a ceramic decoration (Fig. 9) composed of fifteen ochre tiles, the centre representing a flaming sun with a human face of penetrating gaze intensified by a floriated frontal eye. The sun is flanked by two dove-like birds which are perched on a grape-laden vine spiraling the solar orb, the vine leaves formed as five-pointed stars. The bird to the sun's right is concentrated on the rays; the one to its left looks back toward the fruit. The provenance of this mural is a Parisian mansion of the sixteenth or seventeenth century.

The symbolism here is patent: from a Hermetic source we have the same doctrine as that of the 'two birds' in the Vedic passage cited above—Sunbird and Soulbird, the one being the Intellect and the other the individual self. The first is inwardly concentrated on the Light of Reality, on the Supreme Subject (or Object, since they are

1. Tr. Robert Ernest Hume.

identical here); the second is seduced by the fruits of outward, phenomenal existence. Whether the fruit be grapes or apples, it comes to the same. The vine itself, terminating in the frontal eye of the Sun that floats above the Summit of the central trunk, is thus clearly the Tree of Life.

Duo sunt in homine, as Coomaraswamy tirelessly proclaimed; it is for us to gather these two selves together, and of 'the twain make one' (see *Ephesians*, II.15).

'This is not the work of a day', as some authority put it, although from another and Sufic point of view, it is no more than a matter of 'one step—from the shadows to the sunlight'. But a total commitment is required. In the words of Hakuin (1683–1768), the Japanese Rinzai Zen master, 'This immediate attainment of Buddhahood by the one realization is entirely the exertion of one moment of absolute energy.'[1] What this means is elucidated in the Gospel injunction: 'Thou shalt love the Lord thy God with all thy heart, and with all thy soul, and with all thy mind, and with all thy strength.' Everything necessary is contained in this formula, for getting us out of our 'split-gate' existence.

But the truth is, the world of dualities locks most people in its embrace; it is *this* they love with *all their being*, as representing reality, or at least the only reality that makes any 'sense'.

And yet since duality cannot be present without form, and since form cannot manifest without separation, and since separation is but another term for incompletion, how then can all this add up to reality in any permanent, final sense?

The 'split-gate' world, however, is within its limits effectively real inasmuch as its Source wills its existence: 'To show forth his wisdom has He made all things (Hermes, *Lib*.III.la). 'And God saw every thing that he had made, and, behold, it was very good' (*Genesis*, I.31).

Our world is certainly real as a *modality* of Reality, but being a modality it therefore cannot be self-determined. Its apparent 'reality' results from the stability that comes through the 'seamless' interpenetration of polarities, complementaries, and oppositions. This apparent reality can be compared to the colors resulting from the polarization of undifferentiated light when projected through a prism.

1. *The Embossed Tea Kettle and Other Works of Hakuin Zenji*, tr. R.D.M. Shaw, London, Allen & Unwin, 1963, p. 144.

Metaphysically it is the Logos—analogous here to the *candi bentar*—which projects in differentiated mode on the universe the formless Light of Pure Being. These 'colors' in themselves are real enough, but not as coming from themselves. The miracles frequent in earlier ages, when the earth was less 'solidified' and subject-and-object more unified, were in fact an earnest that the outwardly 'seamless' homogeneity of dualities is not of a nature to exclude the irruption down here of higher realities on propitious occasions.

The whole praxis of worship turns on a twofold evidence: that a higher—eventually ultimate—Reality exists, and that our beatitude can only be found therein (the 'supernatural destiny' of Christian theology). In the world of pairs this Reality is necessarily invisible to our pair of eyes, but it can be seen with the single vision of the Eye of the Heart, or what the *Qur'ân* calls the Eye of Certainty (*'ayn al-yaqîn*). The whole challenge is to get subject and object together again. We have in other words to seek the alternative to the alternatives in polarities. As Giordano Bruno taught, God is the unity of all opposites, a Unity without opposites.

*

Coomaraswamy has written extensively on the Sacrifice: both the one *in divinis*, and its reflected counterpart that has to take place down here. There must be a 'slaying' or 'dismemberment' in the worlds above, as we saw in Chapter 2, in order that the worlds below may come into existence. The prototypal slaying has something brutal about it, which the gods are understandably reluctant to undertake. But on their diremption depends our mortal existence; and it is now incumbent upon us—*noblesse oblige*, given the prerogatives of our high human origins—that the Sacrifice be re-enacted—namely reversed—on our part, that we re-member what has been dismembered, and thus gain our salvation and our Peace, through an expiation towards Heaven accomplished here below. At least such has been the worldwide practice in traditional civilizations up until our time. To reject the Sacrifice is to stand in some way outside the cosmic norm and to ally ourselves with unreality.

There is nothing particularly mysterious about all this, being in fact a cyclical reciprocity adumbrated in our every breath. Inspiration-expiration, night and day, winter-summer, death and life—and inversely.

> For that which is born death is certain, and for the dead birth is certain. Therefore grieve not over that which is unavoidable (*Bhagavad-Gîtâ*, II.27).[1]

> Thou bringest forth the living from the dead, and Thou bringest forth the dead from the living (*Qur'ân*, III.27).[2]

As we cannot escape death, and which is moreover for the spiritually prepared not a loss but a repossession, we have every advantage in accepting the necessary with willing submission rather than with heedlessness or revolt. There is moreover an anticipatory dying (*ars morendi*) which can be done right here and now while in the body for those who know the way, and which is described, for one, by the English alchemist Thomas Vaughan (1622–66), in the following manner:

> (The soul) hath many ways to break up her house, but her best is without a disease. This is her mystical walk, an exit only to return. When she takes the air at this door, it is without prejudice to her tenement.[3]

★

In the *Chândogya Upanishad* (VIII.vi.5), the Sun is spoken of as a 'world-door, an entrance for knowers, a stopping for nonknowers'. This mystery of alternatives is the same essentially as the 'split gate', 'Symplegades', *barzakh*, or 'eye of the needle', through which no 'otherness' can penetrate. 'God is not known,' says the pivotal Sufi teacher, Ibn 'Arabî (1165–1240), 'save by His uniting all opposites in the attribution of them to Him'.[4]

Elsewhere he writes: 'Allâh is exempt from all likeness as well as from every rival, contrast or opposition.'[5]

The Islamic chronicler Al Yâfi'î (1299–1367) recounts the advice a sage gave to the great Sufi Dhu'l-Nûn: 'If thou knewest God, thou

1. Tr. Swami Paramananda.
2. Tr. Marmaduke Pickthall.
3. *Anthroposophia Theomagica*; cited by E.J. Langford Garstin, *Theurgy or the Hermetic Practice, A Treatise on Spiritual Alchemy*, London, Rider, 1930, p. 101.
4. *Fuṣûṣu 'l-Hikam*, tr. R.A. Nicholson.
5. *Risâlat al-Aḥadîyah*; cited by Guénon.

wouldst also know the way which leads to him. Abandon differences and divergences.'[1]

In Vedântic terms as formulated by Śankarâchârya (fl. 800 AD), 'Objects that come into being and are capable of being made the objects of Knowledge are as unreal as those known in dream. As duality has no (real) existence Knowledge is eternal and objectless.'[2] Here we have a supreme example of the Scholastic 'rights of the subject'.

From a Chinese Zen source are the words of the Buddhist master Huang Po (d. *c*850 AD): 'The very nature of the Great Way is voidness of opposition'.[3]

Tibetan Buddhism gives us the following by Padma-Sambhava (fl. eighth century AD), the great teacher who introduced Tantric Buddhism from India into his country:

> In its true state, mind is naked, immaculate; not made of anything, being of the Voidness; clear, vacuous, without duality, transparent; timeless, uncompounded, unimpeded, colourless; not realizable as a separate thing, but as the unity of all things, yet not composed of them; of one taste, and transcendent over differentiation.[4]

It is told in the Sahidic Coptic manuscript (*c*140 AD), *The Gospel According to Thomas*, how Jesus compared children being suckled to those who enter the Kingdom. Upon being asked by his disciples for an explanation, Jesus replied:

> When you make the two one, and when you make the inner as the outer and the outer as the inner and the above as the below, and when you make the male and the female into a single one, so that the male will not be male and the female (not) be female . . . then shall you enter (the Kingdom). Log. 22[5]

1. *Rawdh al Rayâhîn*, Cairo, 1889.
2. *Upadeshasâhasrî*, II.ix.7; tr. Swami Jagadânanda.
3. *The Zen Teachings of Huang Po*, tr. John Blofeld, London, Rider, 1958, p. 70.
4. W.Y. Evans-Wentz, *The Tibetan Book of the Great Liberation*, London, Oxford University Press, 1954, p. 229.
5. Trs. A. Guillamont, H.-Ch. Puech, G. Quispel, W. Till, and Yassah 'Abd al-Masîḥ; Leiden, E.J. Brill, London, Collins, 1959.

Giordano Bruno describes the Sundoor thus:

> There is but one centre from which all species issue, as rays from a sun, and to which all species return.[1]

★

> Come you lost Atoms to your Centre draw,
> And *be* the Eternal Mirror that you saw:
> Rays that have wander'd into Darkness wide
> Return, and back into your Sun subside. (Fig. 10)
>
> 'Aṭṭâr[2]

1. Frith, *Life of Bruno*, Trübner & Co., p. 278.
2. Farîd ad-Dîn 'Aṭṭâr (d. 1229): *Manṭiq-aṭ-Ṭayr*, or *The Bird Parliament*, tr. Edward Fitzgerald, Boston, 1899, p. 187.

Suggested Reading

Frithjof Schuon, *From the Divine to the Human*, Bloomington, IN., U.S.A., World Wisdom Books, 1982

——, *Light on the Ancient Worlds*, World Wisdom Books, 1984

——, *Survey of Metaphysics and Esoterism*, World Wisdom Books, 1986

René Guénon, *Man and His Becoming According to the Vedânta*, London, Luzac, 1945

——, *The Reign of Quantity and the Signs of the Times*, Luzac, 1953

——, *Fundamental Symbols: The Universal Language of Sacred Science*, Cambridge, Quinta Essentia, 1995

Ananda K. Coomaraswamy, *Hinduism and Buddhism*, New York, Philosophical Library, 1943

——, *Selected Papers* (2 vols.), Princeton, 1977

Titus Burckhardt, *Alchemy, Science of the Cosmos, Science of the Soul*, Cambridge, Quinta Essentia, 1995

——, 'Cosmology and Modern Science' (in *Mirror of the Intellect, Essays on Traditional Science and Sacred Art*, Cambridge, Quinta Essentia, 1987

Marco Pallis, *Peaks and Lamas*, London, Woburn Press, 1974

Seyyed Hossein Nasr, *Knowledge and the Sacred*, Edinburgh, Edinburgh University Press, 1981

Martin Lings, *The Book of Certainty*, Cambridge, Islamic Texts Society, 1992

Lord Northbourne, *Looking Back on Progress*, London, Perennial Books, 1970

Philip Sherrard, *Human Image: World Image—The Death and Resurrection of Sacred Cosmology*, Ipswich, Golgonooza Press, 1992
Douglas Dewar, *The Transformist Illusion*, Murfreesboro, TN., U.S.A., Dehoff Publications, 1957
Mircea Eliade, *The Myth of the Eternal Return*, Princeton, 1971
Jean Servier, *L'Homme et l'Invisible*, Paris, Robert Laffont, 1964
Herbert Dingle, *Science at the Crossroads*, London, Martin Brian & O'Keeffe, 1972

Index

Abel, 32, 34
Abélard, Peter, 63
abiogenesis, 4
Absolute, the, 3, 15, 88
Acta Sanctorum, 46n., 47
Adler, Mortimer J., 6
Advaita Vedânta, 14
alchemists, 40, 77
Alexander the Great, 32
Alhambra, the, 12n.
al-Khiḍr, 48
All-Possibility, the, 17
Al-Yâfi'î, 99
Americas, the two, 36
Amidism, 48
amino acid constituents, 4
Amitâbha, 48
analogy, laws of, 65
Angelism, 67n.
Anglican Church, 34
Anima Mundi (World Soul), 14, 92
animism, 38–9
Anthony, St., 46
ao (light), 34
apophatic theology, 14
Aquinas, St. Thomas, 30, 40, 63
archebiosis, 4
Archetypes, 15, 40, 46, 48, 50, 59, 71, 89
Aristotelianism, 63, 69
Ark, the, 18n.
Armitage, Angus, 78n.
ars moriendi, 99
Astrology, 83, 93
astrophysics, 90
Atlantis, 84
'Aṭṭâr' Farîd ad-Dîn, 101
Avalokiteśvara, 48–50
Axis Mundi, 27, n., see axis (world)
axis (world), 27, 41, 48, 75–6

Bacon, Francis, 66–7
Bacon, Roger, 65–6
Bali, 11, 12, 38ff
Bali Aga, 38f
Balinese, 38ff
Barbara, St., 46
Barber, Noel, 62n.
Bari, 47
Barmont, Louis, 74
Barr, James, 93n.
barzakh, 12, 13, 24
Beatific Vision, 27–8
becoming, 88
Bedouins, 32
Being, 69, 88, 98
Bell, John Stuart, 91
Bell's Interconnectedness Theorem, 91ff
Berengarius of Tours, 60
Berkeley, George, 69–70, 91
Besakih, 41

Big Bang, 16, 86
Bhagavad-Gîtâ, 99
bifurcation concept, 64
Bligh, Captain William, 34
Bodhisattvas, 48
Boehme, Jacob, 16, 27
Bolivar, Simon, 32
Bollandists, the, 46n.
Boorstin, Daniel, 50, 77, 79, 82
Bounty, HMS, 34
Boyle, Robert, 67, 79n.
Brahe, Tycho, 78–9
Brahmanism, Shaivite, 38
Bṛihad Âraṇyaka Upanishad, vii
Bruno, Giordano, 77–8, 80, 98, 101
Bryce, Viscount, 61
Buddha, 49
Buddhism, 38, 56
Burckhardt, Titus, 64
Byzantium, 60

Caesar, Julius, 32
Caesar (function), 57–8
Cain, 32, 34
carbon compounds, 4
Catherine of Alexandria, St., 46–7
Catherine of Siena, St., 56n.
Center of the World, 41, 51
Chândogya Upanishad, 99
Charlemagne, 57, 59
Chartres Cathedral, 47
Chivalry, 56
Christianity, 37, 55–6
Christmas Tree, 26–7, 47
Christopher, St., 46, 75n.
chronology, Balinese, 40–41
Chuang-tse, 6
Church of Christ, Scientist, 70
cognitive pleroma, vii, 10
Coleridge, Samuel Taylor, ix, n., 7
Columbus, Christopher, 52
Constantine the Great, 50, 60ff
Constantinople, 46, 60–61
Cook, Captain James, 35
Cook, Roger, 27n.
Coomaraswamy, Ananda, x, 8n., 12, and n., 15, 17n., 32, 98
Copernicus, Nicholaus, 50, 77ff
Cosmas of Alexandria, 50
cosmic cycle, 23–4
Coulter, Harris L., 93n.
'counterintuitive' evidences, 93n.
Covarrubias, Miguel, 11, 12n.
Creationism, x
Crusades, 62
cults, 90
Cutsinger, James S., ix, n.
cyclical outlook, 44

Dafydd ap Gwilym, 22
Dante, Alighieri, 9, 22, 57–8, 61, 75
Dark Ages, the, 45, 53, 58–9
Darwin, Charles, Darwinism, ix, 67, 87
Day of Reckoning, 84
Descartes, René, 67
devolution, 6
Dhu'l-Nûn, 99
Diocletian, 46
Dionysius, 14
Dodd, Edward, 33n.
Donation of Constantine, 61
Donne, John, 81–2
Doxopatrius, Nilus, 58
Dürer, Albrecht, 73ff
Dutch administration (Bali), 42–3
dust, primal cosmic, 4

Ecclesiasticus, 1, 9
Eckhart, Johannes, Meister, 9, 14, 19
'Ecumenical' maps, 51
Eddington, Sir Arthur, 69, 70n.
Einstein, Albert, 77, 88, 90–91, 92n.
Eliade, Mircea, 45
emanationism, 86
Emerald Tablet, the, 42
Emerson, Ralph Waldo, 3
Engels, Friedrich, 87
Epigoni, 32
equal temperament, 78n.
Eratosthenes, 50
Erigena, Johannes Scotus, 14, 63, 83
ether, 92
Eucharist, 60
Everest, Mt., 6

INDEX

Evil, 19
eye-inner, frontal, single, 53, 96, 98

Fall, the, 27
fallen man, 34
Ferdinand and Isabella, 52n.
filioque, 60
flat-earth perspective, 50, 53
flat-world syndrome, 55
Flood story, 18n.
Forman, Bedřich, 11n.
Forman, Werner, 11n.
formless manifestation, 12, 15
Foster, David, 64–5n.
Frederick II, 58
Freud, Sigmund, 88–9

Galileo, 67, 77, 82
Garden of Eden, 52
Genesis, 8
geocentric system, 50, and passim
George, St., 46–7
gnomonic projection, 54
Golden Age, 28
good and evil, 17, 18, 26
Gospel According to Thomas, The, 100
Gospel injunction, 97
gravity, 16, 66
Great Spirit, 36–7
Gregory VII, Pope, 62
Gregory IX, Pope, 58n.
Guénon, René, x, 16, 18, 19, 22, 24, 31, 32, 65, 75
gunas, Hindu, 25
Gunung Agung, 41
Gurdjieff, G. I., 56

Hagia Sophia, 60, 62
Hakuin, 97
Harvey, William, 67, 79n.
Hawaiki, 33
heliocentric system, 50, and passim
Herbert, Nick, 90n., 91n.
Hering's Law, 30n.
Hermes, 22, 42, 93, 97
Hermeticism, 56
Herrick, Robert, 16
Herrigel, Eugen, 93n.
Higher Waters, 12
Hinduism, 38, 55–7
Hipparchus of Nicaea, 50
historicity, secular, 44
Hobbes, Thomas, 67
Holy Ghost, 48
Holy Sepulchre, Church of the, 52, 58n.
homoeopathic medicine, 30n.
hourglass, 24
Howarth, David, 33n., 34
Huang Po, 100
Humbert, Cardinal, 60
Hume, David, 70

Ibn 'Arabî, 99
Ibn Idrisi, 58
Ideas, 59, 68
Indians (American), 36
Indonesians, 38
Infinite, the, 3, 15
Intellect, the, 13, 53, faculty, 71–2, 96
intelligence, artificial, 6n.
inverse analogy, law of, 24
'Irâqî, 22
Iron Age, 34
Ironquois, 37
Irving, Washington, 52n.
Islam, 13n., 38, 45, 55–7

Jâmî, 70
Jewry, mediaeval, 47
John the Baptist, St., 75
John of Damascus, St., 48
John the Evangelist, St., 75
Jones, Charles W., 48n.
Judgment Day, 13n.
Jung, Carl, 89
Justinian the Great, 46

Kabbalists, 58
Kali-Yuga, 34, 36, 40
Kalki Avatar, 76
Kant, Immanuel, 71–2
Kent, Dr. James Tyler, 30n.
Kepler, Johannes, 79, 81
Kinross, Lord, 62n.
Kwannon, 49

Kwannon Sutra, 49
Kyffhäusser Mountain, 59

Lactantius, Firmianus, 50
Lamarck, Jean Baptiste, 87
Laux, John, 62n.
Law of Corollaries, 3
Law of Dualities, 3
Law of Opposites, 3
Law of Pairs, 3
Law of Polarities, 3, 7, 13
Leibniz, Gottfried Wilhelm, 68
Leo IX, Pope, 60
Lesser Mysteries, 76
light, speed of, 89n.
Lîlâ, 17
Locke, John, 68
Logos, 25, 98
London Missionary Society, 34
Lower Waters, 12, 41

Macaulay, Lord, 54
Mach, Ernst, 78n.
Magellan, Ferdinand, 32
Mahameru, 41
Mahayana Buddhism, 14
Mahdi, the, 76
Maitreya Buddha, 76
Majâpahit dynasty, Hindu, 39
Malayans, Proto-, 38
Manichaeism, 18
Manifest Destiny, 36
Maritain, Jacques, 67n.
Martin, Malachi, 10n., 62n.
Marx, Karl, 87
Materialism, 35
materia prima, 7, 25, 40, 69, 92
Mauna Kea, Mt., 6
Mâyâ, 7, 16, 23
Mediaeval Europe, 37, 50, 87
Melanesians, 38
Melchizidek, 48
'Melencolia', 73ff
Mercator projection, 54
Meru, Mount, 54
Michelson-Morley experiment, 77
Middle Ages, 63
Migne's *Patrologia Graeca*, 48
Migne's *Prologia Latina*, 48
Milton, John, 82
miracles, 98
missionaries, 36, 42
Mitra-Varuṇa, 8
Mongols, 35
Morse, Samuel, 32
Motionless Mover, 14
Mrázek, Rudolf, 11, 11n.
Myra, 47
myroblite, 47

nâma-rûpa, 21
Napoleon Bonaparte, 32
Native Americans, 36
Navel of the world, 41, 47
navigation, Polynesian, 32f, 93
Neoplatonists, 80
Newtonian universe, ix, 88
Newton, Sir Isaac, 79, 90
Nicene Creed, 60
Nicholas, St., 46ff, 58, 94
Nicholas of Cusa, 61, 80
Nicholas of Flue, St., 56n.
nomads, 32ff
Nominalism, 59–60, 63
Non-acting Action, 19
nothingness, absolute, 6
noumena, 71

object, 1ff, and *passim*
Ockham, William of, 59, 63, 65–6
Ockham's Razor, 64
Oecumenical Council of Nicea, first, 46
one-world syndrome, 55
Ormazd, 18
Orthodox Church, the, 44–5
Ottoman Empire, 62

Padma-Sambhava, 100
Pañcavimśa Brâhmaṇa, 8
panentheism, 21
Paracelsus, 8n., 93n.
Parmenides, 83
Pascal, Blaise, 8, 67
pawnbrokers, Lombard, 47
Perry, Whitall N., 30n., 46n., 92n.

INDEX

Peter Damian, St., 48
Pillars of Hercules, 53
Plains Indians, 36
Plato, Platonism, 27, 59, 63
Platonic year, 83
Platonists, Cambridge, 68
Plotinus, 83
po (darkness), 34
Polo, Marco, 52
Polynesians, 32f, 38, 41, 93
Powers of Darkness, 18, 89
Powers of Light, 18
Prakriti, 7, 15, 21
pralâyâ, 16
Praśna Upanishad, 10
Precursor, the, 76
primordial man, 26, 31, 54, 83, 89
Primordial State, Tradition, 28, 36, 75, 85
protoplasm, 4
protozoa, 4
psychic influences, 39
Ptolemy, 50, 53, 78ff, 94n.
Puerta Judiciaria, 12n.
puputan, 42
Purusha, 7, 15, 21
putrefaction, alchemical, 73
Pythagorean system, 65, 78n.

quadrivium, 65
qualitative pole, 9, 65, 66
quantitative pole, 9, 65, 66
quantum physics, 86, 90ff
Qur'ân, 1, 9, 11, 12, 16, 17, 25, 26, 43, 80, 93, 99

Ratramnus, 60
Ravenna, 46
Realism, 59ff
Recollection, intellectual, 68, 90
relativism, 88
religion, Indian (American), 37
Renaissance, 37
restoration (cosmic), 75
reversal of the poles, 24, 74–5, 89
Roger the Norman, 57
Roscellinus of Compiègne, 59
Round of Existence (*saṁsâra*), 21
Royal Society, 35

sacrifice, 16, 98
Śakyamuni, the Buddha, 48
saṁsâra (Round of Existence), 21, 80
Śankarâchârya, 100
San Marco, Basilica of, 46
S. Apollinare Nuovo, 46
Saracens, 47
Satan, reign of, 74
Sat-Chit-Ànanda, 15
Schism, Great, the, 60
Scholasticism, 59, 63, 69
Schroedinger, Erwin, 64n.
Schuon, Frithjof, x, 5, 7, 13, 17, 31, 53, 71, 89n.
science, modern, 37, 64, 84
science versus religion, ix
Scupoli, Lorenzo, 14
sedentaries, 32ff, 53, 55, 88
Shabistarî, Sa'd al-Dîn Maḥmûd, 8
Shakespeare, William, 77
Shakti, 47, 74
shamanistic vision, 37, 93
Sherrard, Philip, 10n.
Smith, Huston, 93n.
Smith, John, the Platonist, 63
Smith, Wolfgang, 10n., 77–8
solstice, summer/winter, 75–6
space, 24, 32f
space, six directions of, 36
Spencer, Herbert, 87
Spenser, Edmund, 81
Spinoza, Baruch, 68
spiritual authority, 57
Spiritus, 7
split gate, 11–13, 24, 27, 83, 97, 99
Sterry, Peter, 19
subject, 1ff, and *passim*
Substance, Divine, 7
subtle prototypes, 40
subtle state, forces, order, 39, 89, 91–2
Sufic saying, 97
Suhrawardi, 63
Supernal Sun, 27, 30, 81
supernatural destiny, 98

Supreme Being, 34
Supreme Center, 28, 81
Supreme Principle, 14, 48
Suzuki, D.T., 49n., 84n.
S. Vitale, 46
Sylvester I, Pope, 61
Sylvester II, Pope, 58n.
symbolism, 47
Symplegades, 12, 13

Taaroa, 33–4
taboos, 35
Tahiti, 34
Tahitians, 33ff
Takuan, 84
tantra, 56, 100
Tartars, 32
Teilhard de Chardin, Pierre, 10
temporal power, 57
Ten Commandments, the, 34–5
Terrestrial Paradise, 27n., 52
Thermodynamics, Second Law of, 70n.
throne, Queen of Sheba's, 93
Tilak, B.G., 27n.
time (linear, cyclical), 24, 32f, 86, 94
'Titanic', 88
'T-O' maps, 51
Torcello, 46
tortoise, symbolism of, 54
Toscanelli, Paolo del Pozzo, 52
transformism, ix, 1, 26, 40, 46n.
Tree of Life/Knowledge/World Tree, 25, 27, 97
Trinitarian doctrine, Christian, 15
Tritheism, 60
trivium, 65

Two Rivers, Johnny, 37

Universal Man, 25, 83
Universals, 59
Upanishads, 83
Urban II, Pope, 62
Ursula, St., 46

Valla, Laurentius, 61
Vatican II, 60, 91
Vaughan, Thomas, 99
Vedânta, Advaita, 70
Venice, 46
vicegerent (man as), 25, 43, 54, 74
Vikings, 32f, 57
Virgin Mary, 13n., 47
viruses, filter-passing, 4
Vithoulkas, George, 30n.
Voltaire, 68
von Neumann, John, 91

Walker, Benjamin, 54n., 55n.
Ward, John, 93n.
Warren Court, 91
Warren, William F., 27n.
Wheeler, John, 91
Whitehead, Alfred North, 64
Wilber, Ken, 64–5n.
worship, 98

Yab-Yum, Tibetan Tantric, 15, 23
yin-yang, 22–4, 28

Zarlino, Gioseffo, 78n.
Zen archery, 93; Formula, 14; koan, 37
Zodiac, 31, 82n., 93
Zoroastrianism, 18
Zrvana Akarana, 18

BY THE SAME AUTHOR

A Treasury of Traditional Wisdom, Allen & Unwin, and Simon & Schuster, 1971
Second edition, Perennial Books, 1981
Third edition, Harper & Row, 1986
Fourth edition, Quinta Essentia, 1991

Gurdjieff in the Light of Tradition, Perennial Books, 1978
Second impression, 1979

Challenges to a Secular Society, Foundation for Traditional Studies, in preparation.